Canada's Arctic waters in international law

Studies in Polar Research
This series of publications reflects the growth of research activity in and about the polar regions, and provides a means of disseminating the results. Coverage is international and interdisciplinary: the books will be relatively short (about 200 pages), but fully illustrated. Most will be surveys of the present state of knowledge in a given subject rather than research reports, conference proceedings or collected papers. The scope of the series is wide and will include studies in all the biological, physical and social sciences.

Editorial Board
R. J. Adie, British Antarctic Survey, Cambridge
T. E. Armstrong, Scott Polar Research Institute, Cambridge
D. J. Drewry, Scott Polar Research Institute, Cambridge
B. Stonehouse, Scott Polar Research Institute, Cambridge
P. Wadhams, Scott Polar Research Institute, Cambridge
D. W. Walton, British Antarctic Survey, Cambridge
I. Whittaker, Department of Anthropology, Simon Fraser University, British Columbia

Other titles in this series:
The Antarctic Circumpolar Ocean
Sir George Deacon
The Living Tundra
Yu. I. Chernov, transl. D. Love
Transit Management in the Northwest Passage
edited by C. Lamson and D. Vanderzwaag
Arctic Air Pollution
edited by B. Stonehouse
The Antarctic Treaty Regime
Edited by Gillian D. Triggs

Canada's Arctic waters in international law

DONAT PHARAND, Q.C., F.R.S.C.
Professor of Law, University of Ottawa

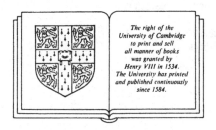

The right of the
University of Cambridge
to print and sell
all manner of books
was granted by
Henry VIII in 1534.
The University has printed
and published continuously
since 1584.

CAMBRIDGE UNIVERSITY PRESS

Cambridge

New York   New Rochelle   Melbourne   Sydney

CAMBRIDGE UNIVERSITY PRESS
Cambridge, New York, Melbourne, Madrid, Cape Town, Singapore, São Paulo, Delhi

Cambridge University Press
The Edinburgh Building, Cambridge CB2 8RU, UK

Published in the United States of America by Cambridge University Press, New York

www.cambridge.org
Information on this title: www.cambridge.org/9780521325035

© Cambridge University Press 1988

First published 1988
This digitally printed version 2008

A catalogue record for this publication is available from the British Library

Library of Congress Cataloguing in Publication data
Pharand, Donat
Canada's Arctic waters in international law
(Studies in polar research)
Bibliography
Includes index.
1. Territorial waters – Northwest Territories – Arctic Archipelago.
2. Maritime law – Canada.   3. Territorial waters.   4. Maritime law.
I. Title.   II. Series.
JX4131.P48   1987   341.4'48'097199   86-26395

ISBN 978-0-521-32503-5 hardback
ISBN 978-0-521-10006-9 paperback

To MICHEL
    and
    BERNARD
    and
    GISÈLE

Contents

Contents vii

Figures and Tables

Foreword

In 1983 Professor Donat Pharand spent a six-month study leave with Dalhousie Ocean Studies Programme in Halifax, Nova Scotia. For many of us at D.O.S.P., and also at Dalhousie Law School, this was a period of reunion and reacquaintance with an old friend and fellow alumnus. For students it was an introduction to one of Canada's most dynamic law teachers. For me it is a special pleasure to be asked to supply the foreword to the product of these years of labour.

There must be few readers indeed who are unaware of Professor Pharand's work and reputation in the specialized area of Arctic Ocean law and policy. His earlier book *The Law of the Sea of the Arctic* (University of Ottawa Press, 1973) is, of course, still one of the standard reference works in the area, but the general law of the sea has undergone major surgery in recent years, and the policy issues associated with the Arctic Ocean have assumed a more critical significance as a result of more pressing resource and industrial demands on this remote and unique environment. One of a series of new studies by Professor Pharand, this monograph re-examines in detail the theories and practices legally relevant to the waters of the Canadian Arctic Archipelago, in light of the most up-to-date research data as well as the most relevant portions of the historical record. Scrupulously documented and strongly argued, it presents the same erudition and rigour as its predecessors.

These and related Arctic issues in the law of the sea are of major interest to us at D.O.S.P. The future use of the Northwest Passage is a matter of serious and continuing concern to most Canadians. Accordingly, late in 1982, we initiated an ambitious cross-disciplinary programme of studies on the problems of 'transit management' in that region with a major four-year grant from the Donner Canadian Foundation. This work is the second of our own series of monographs arising from the

Canadian Northern Waters Project. At the time of writing, the transit of the Passage by a U.S. Coast Guard vessel has revived the full range of sentiments and arguments aroused by the *Manhattan* crossing of 1970. Whatever one's views on these legal issues, the timeliness of the transit management debate is, once again, evident to all. We welcome this latest contribution to our series, and predict another scholarly success for its distinguished author.

Halifax, Nova Scotia

Douglas M. Johnston,
Director, Canadian Northern Waters Project,
Dalhousie Ocean Studies Programme

Preface

It is presently an opportune time to re-examine the legal regime applicable to the waters of the Canadian Arctic Archipelago. Although it is probable that Arctic hydrocarbon production and transportation will not occur before the late 1990s, the current lull in exploration activity makes appropriate planning possible.

The Arctic Pilot Project – a proposal to ship liquefied natural gas by tanker from Melville Island through the Northwest Passage to a southern Canadian port – has been rejected because of insufficient information as to markets. American developers in the Beaufort Sea have not yet found sufficient commercial reserves to warrant tanker transportation through the Northwest Passage to the American eastboard. Canadian developers – Dome, Esso and Gulf – are still in the process of assessing hydrocarbon reserves in the Beaufort Sea and, despite satisfactory results of certain well tests in 1984, commercial production has yet to begin.

In addition to the above, the Beaufort Sea Environmental Assessment Panel has recommended that government adopt a phased approach to hydrocarbon production and transportation, and has expressed a preference for initial transport of oil by a small-diameter pipeline.[1] However, the Panel recognized that 'certain factors may make a phased approach beginning with tanker transportation the favoured mode of oil transport.'[2] Indeed, in 1983, a special Senate Committee had already seen more advantages in starting with a small tanker system – particularly its flexibility to adapt to reserve levels and market demands – and had recommended that mode to the Government.[3]

In these circumstances, the Beaufort Sea Panel also recommended that the Government approve the use of oil tankers subject to certain conditions. Because of the potential effect of year-round tanker traffic on the biological and physical environment of the waters, the approval would be given only if 'a comprehensive Government Research and Preparation Stage is completed by governments and industry; and a Two-Tanker stage using Class 10 oil-carrying tankers demonstrates that environmental and socio–economic effects are within acceptable limits.'[4] Since these conditions would take several years to fulfil, the Government was urged to proceed immediately with the construction of an icebreaker which would meet at least Arctic Class 8 specifications to carry out some of the necessary studies and eventually support Arctic shipping.[5] Indeed, a Class 10 would be necessary if Canada wishes to provide adequate ice-breaking services and exercise control over *all* of the waters of its Arctic Archipelago. And it is not surprising that, upon being told that 'there are problems in the overall control and monitoring of ship vessel traffic through the Canadian Arctic' the Panel reminded the Government that, 'if a vessel is disabled or sinking in Arctic waters, or has encountered some other emergency, government's responsibility must be absolute and its actions must be swift and unencumbered by jurisdictional or communications problems.'[6] To this end the Panel recommended that 'the present vessel traffic management system, NORDREG, be made mandatory for all vessels which enter Canadian Arctic waters' and the system be extended to the Beaufort Sea.[7] There can be no doubt that such a comprehensive vessel traffic management system is absolutely necessary for the protection of the sensitive marine environment of the Arctic and, now that it has been made clear that the waters of the Arctic Archipelago are internal waters of Canada, the recommendations should be implemented.

The final clarification made by Canada as to the precise status and extent of the waters of its Archipelago was made on September 10, 1985, by the establishment of straight baselines from which Canada's 12-mile territorial sea now extends. The announcement specified that the newly established baselines defined 'the outer limit of Canada's historic internal waters'.[8] Presumably, this means that Canada claims to have acquired an historic title to those waters by the exercise of exclusive authority for a long period of time. If so, it is possible that it might also be relying on its long-standing practice of showing on its official maps its Arctic boundaries as extending to the North Pole, following the 141st and 60th meridians of longitude pursuant to what is commonly known as the 'sector theory'. Any of those legal concepts could conceivably be invoked, either separately or in combination, as legal bases for the international validity for the straight baselines around the Canadian Arctic Archipelago. This study examines these three possible legal bases: the sector theory, the doctrine of historic waters, and the straight baseline system itself. These constitute the first three parts of the book. The fourth and concluding part is devoted to an appraisal of the legal regime applicable to the Northwest Passage. This part reviews a certain number of measures which Canada has either taken already or could take in order to exercise more control over the waters of the Passage.

Ottawa, Canada

Donat Pharand

Notes to Preface

1. See Final Report of the Environmental Assessment Panel, *Beaufort Sea Hydrocarbon Production and Transportation* (July 1984), at 71.
2. *Ibid.*, at 5.
3. Report of the Special Committee on the Northern Pipeline, *Marching to the Beat of the Same Drum: Transportation of Petroleum and Natural Gas North of 60°* (March 1983), at 4.
4. *Supra*, note 1, at 70.
5. *Ibid.*, at 97.
6. *Ibid.*, at 96.
7. *Ibid.*, at 97.
8. Statement in the House of Commons by Secretary of State for External Affairs Joe Clark, 10 Sept. 1985, reproduced in *Statement* Series 85/49, at 3.

Acknowledgements

This book was written at the request of the directors of the Dalhousie Ocean Studies Programme and as part of the Canadian Arctic Waters Project within that Programme. During my stay with D.O.S.P. in Halifax, as senior research scholar for the first half of 1983, I benefited from the wise counsel of Professor Douglas M. Johnston, the Project Director, and received assistance from research associates David VanderZwaag and Cynthia Lamson. In the latter part of 1985, when a revision was rendered necessary because of the adoption of certain legislative measures by Canada, I also received indispensable assistance from Dongdong Huang, a doctoral student at the Faculty of Law of the University of Ottawa. I am very grateful to all those people for their valuable contribution.

A second group of people to whom I gladly acknowledge my gratitude are those who contributed to the preparation of the maps and diagrams relating to the Northwest Passage: S. B. MacPhee, the Dominion hydrographer; Richard Cashen of the Canadian Hydrographic Service, and Commander John Cooper, consultant on maritime boundary delimitation. I am also obliged to Leonard H. Legault, Legal Adviser of the Department of External Affairs, for his permission to draw upon a study I did there in 1979, during my year as academic-in-residence. In addition, I am indebted to the personnel of the Scott Polar Research Institute of the University of Cambridge, particularly Dr Terrence Armstrong, who were most helpful during my three-month research period at that Institute.

The bulk of this study was completed during my sabbatical year of 1983–4 and I am pleased to acknowledge the grant of a fellowship, during that time, from the Social Sciences and Humanities Research Council of Canada. I am also grateful to the authorities of the University of Ottawa for the award of a sabbatical year and to Dean Raymond A. Landry, of the Civil Law Section of the Faculty of Law, for his encouragement and support.

A special word of appreciation goes to my son Michel, an instructor in the Comparative Literature Program at the Pennsylvania State University, who read all of the manuscript and made a number of suggestions (most of which I accepted) to improve the form and style. I am also greatly indebted to Professor D. M. McRae, of the University of British Columbia Law School, who took the time to read the manuscript and to make most valuable comments and suggestions.

Three persons contributed mainly to the typing of the manuscript, most of the time under pressure to meet deadlines; they are Lise Fraser, in Ottawa, and Ena Morris and Lynda Corkum, in Halifax. I am deeply grateful for their indispensable assistance.

Finally, I wish to express my appreciation to Martinus Nijhoff Publishers who permitted me to draw upon a book they published in 1984, *The Northwest Passage: Arctic Straits,* and which necessarily overlaps with Part IV of the present book. My wife Yolaine also deserves a very special word of gratitude; without her moral support and understanding, particularly during my rather long research periods away from home, this book would never have been completed.

Part 1
The waters of the Canadian Arctic Archipelago and the sector theory

The sector theory has been invoked by a number of politicians and officials in Canada as a legal basis for claiming jurisdiction not only over the islands of the Canadian Arctic Archipelago, but also over the waters within and north of the islands right up to the Pole. However, the government itself has never taken a very clear and consistent position on this theory. It would seem that present government policy is to hold the theory in reserve as possible support for its claim that the waters of the Archipelago are internal.

The purpose of this first Part is to assess the validity of the sector theory in international law as a basis for claiming jurisdiction to Arctic waters. This will be done in four chapters, presenting: 1) a brief inquiry into the origins of the theory; 2) a study of the relevant boundary treaties; 3) an analysis of the related concept of contiguity; and 4) a review of State practice and its possible acceptance as customary law.

1

The origins of the sector theory

The paternity of the sector theory is generally attributed to Senator Pascal Poirier who invoked it in 1907 as a basis for claiming sovereignty over all of the islands north of Canada. This attribution is accurate only in so far as he was the first to actually systematize the use of meridians of longitude to claim territorial sovereignty in the Arctic. The use of such meridians had been advanced long before to mark the delimitation of territorial claims.

1.1.   Papal Bull (1493) and early treaties (1494 and 1529)

On May 4, 1493, Pope Alexander VI issued his Bull *Inter Caetera* whereby he purported to grant to Spain 'all islands and mainlands . . . towards the west and south, by drawing and establishing a line from the Arctic pole . . . to the Antarctic pole . . ., the said line to be distant one hundred leagues towards the west and south from any of the islands commonly known as the Azores and Cape Verde'.[1] The Bull safeguarded previous grants to Portugal with a proviso that no similar right already conferred to a Christian prince was to be withdrawn.

On June 7 the following year, Spain and Portugal concluded the Treaty of Tordesillas in which they used a meridian of longitude from Pole to Pole to establish a boundary which alloted Spain 'all lands, both islands and mainlands' on the west side of the line and those on the east side to Portugal. The dividing line was fixed at a distance of 370 leagues west of the Cape Verde Islands, or in other words 270 leagues farther west than established by the Pope in the Bull.[2]

By the Treaty of Saragossa of April 22, 1529, the same Parties agreed to use another meridian of longitude from Pole to Pole, on the other side of the globe, $297\frac{1}{2}$ leagues east of Molucca Islands as a delimitation line for their respective claims.[3]

1.2.   Joint address from Canada to Great Britain (1878)

The first indirect use of the sector theory seems to have been made on May 3, 1878. On that date, the House of Commons and Senate of Canada adopted a joint address to the British Parliament asking for the transfer of all Arctic lands and islands lying between the 141st meridian of longitude and the series of straits between Ellesmere Island and Greenland. The reason for the address was the uncertainty of the extent of the lands and territories transferred to Canada by Great Britain in 1870. The Imperial Order in Council of June 23, 1870 had transferred 'all lands and Territories within Rupert's Land',[4] originally granted to the Hudson's Bay Company in 1670, and it was generally agreed that Rupert's Land comprised only the lands and territories which drained into Hudson Bay and Hudson Strait.[5] Consequently, the British Government felt it advisable to make an additional and more complete transfer to Canada.

The description of the lands and territories asked for in the joint address read as follows:

> . . . on the East by the Atlantic Ocean, which boundary shall extend towards the North by Davis Straits, Baffin's Bay, Smith's Straits and Kennedy Channel, including all the islands in and adjacent thereto, which belong to Great Britain by right of discovery or otherwise; on the North the Boundary shall be so extended as to include the entire continent to the Arctic Ocean, and all the islands in the same *westward to the one hundred and forty-first meridian west of Greenwich;* and on the North-West by the United States Territory of Alaska.[6]

Great Britain, however, was not willing to take the chance of purporting to transfer more lands than it had title to and, therefore, limited its transfer of 1880 to 'all British Territories and Possessions in North America, not already included within the Dominion of Canada, and all Islands adjacent to any such Territories or Possessions . . . (with the exception of the Colony of Newfoundland and its dependencies) . . .'.[7]

In these circumstances, considerable doubt remained as to the exact legal status of some of the islands north of Canada, and the government and its officials began to take a number of steps to consolidate Canada's claim to those islands.

1.3.   Description of Franklin District (1897)

On October 2, 1895, on the basis of a report prepared by the Minister of the Interior, an Order in Council (P.C. 2640) was adopted which divided the unorganized Northwest Territories into four provisional districts: Ungava, Franklin, MacKenzie and Yukon. The western boundary of Franklin, the northernmost district, was described as beginning at about 125°30′ longitude west and reaching a latitude of about 83¼° north, and the Order in Council confines itself to mentioning three islands: Baring Land (what is now the southern part of Banks Island), Prince Patrick Island and the Polynea Islands.

The above Order in Council, considered defective in its description of the Franklin District, was annulled by a subsequent Order in Council dated December 18, 1897 (P.C. 3388), which remedied the situation by describing the Franklin District as follows:

> The District of Franklin (situated inside of the grey border on the map herewith) comprising Melville and Boothia Peninsulas, Baffin, North Devon, Ellesmere, Grant, North Somerset, Prince of Wales, Victoria, Wollaston, Prince Albert and Banks Lands, the Parry Islands and all those lands and islands comprised between the one hundred and forty-first meridian of longitude west of Greenwich on the west and Davis Strait, Baffin Bay, Smith Sound, Kennedy Channel and Robeson Channel on the east which are not included in any other Provisional District.[8]

Although the North Pole is not mentioned, this description of the provisional district of Franklin, by reference to the 141st meridian on the west and the Davis Strait, Baffin Bay, South Sound, Kennedy Channel and Robeson Channel on the east, incorporates a claim to 'all lands and islands' contained within the sector constituted by the *141st and 60th meridians.* This is made quite clear by an examination of the map prepared by J. Johnston, geographer of the Interior. The map was attached to the Order in Council and incorporated by reference in the description (see Figure 1).[9]

It should be noted, however, that this map did not extend the boundaries right up to the North Pole. The boundary line along the 141st meridian extends to 78½′ of latitude and the line along the 60th, up to the 85th. The boundaries were not shown as extending to the Pole, since the meridians followed did not extend any farther north on the map used.

1.4.   Maps of explorations (1904) and territorial divisions (1906)

In 1904, the Department of the Interior published a map entitled 'Explorations in Northern Canada and Adjacent Portions of Greenland and Alaska', which showed *the 141st and 60th meridians, extending right up to the North Pole,* as being Canada's boundaries (see Figure 2). The map lists the explorations of coasts and indicates the tracks of the various

Fig. 1. Provisional Districts of Canada, 1897, *Source:* Public Archives of Canada.

Fig. 2. Explorations in Northern Canada, 1904, *Source*: Public Archives of Canada.

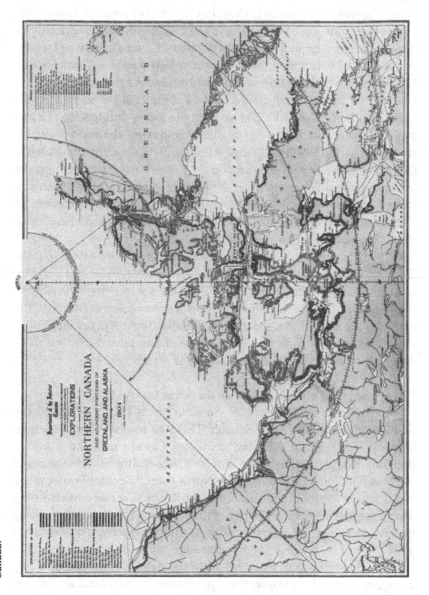

northern expeditions. It was prepared by the departmental geographer, James White, and incorporates the names of the minister and deputy minister of the Department of the Interior.

The 1904 map was published at the time of the expeditions of A. P. Low, of the Geological Survey of Canada, who had been sent, in August 1903, in command of the *Neptune,* to patrol the waters of Hudson Bay and those adjacent to the Arctic islands. In the words of Andrew Taylor, 'the primary purpose of the voyage was to plant the flag in numerous localities throughout the Arctic; it was the signature of Canada's growing interest in the islands, sovereignty over some of which was being contested by Norway on the basis of the work of Sverdrup and Amundsen'.[10]

It will be recalled that Sverdrup had completed his second Arctic expedition from 1898 to 1902 in his ship, the *Fram,* during which he had explored the northernmost Arctic islands, and Amundsen had begun his voyage across the Northwest Passage in June 1903, which he completed in October 1906. An additional reason for Low's voyage was perhaps the fear of a United States claim on the basis of explorations by American nationals. Perry in particular had explored the east and north coasts of Ellesmere Island between 1898 and 1902 in order to prepare a base for an eventual expedition to the North Pole.

In 1906, the year Captain Bernier led his first expedition (1906–07), the Department of the Interior published a map entitled 'Atlas of Canada No. 1, Territorial Divisions' (see Figure 3). This was the first official map showing all of Canada, including the northern boundaries. Although the map does not show the boundaries as extending to the North Pole, it does show them as following the *141st and 60th meridians* as far north as necessary to include all of the northernmost islands.

1.5.   Senator Poirier's motion on the sector theory (1907)

In 1907, on February 20, Senator Pascal Poirier proposed the now famous resolution in the Senate: 'That it be resolved that the Senate is of opinion that the time has come for Canada to make a formal declaration of possession of the *lands and islands* situated in the north of the Dominion, and extending to the North Pole'.[11] Senator Poirier made a long speech in support of his motion, which was never seconded, during which he quoted from the Charter of the Hudson's Bay Company, saying that he had seen old maps showing territories discovered and taken possession of by English sailors, extending up to and above the 82nd parallel. He stated that, as successor to the rights of the Hudson's Bay Company, Canada could now claim '*possession of all the lands and islands*

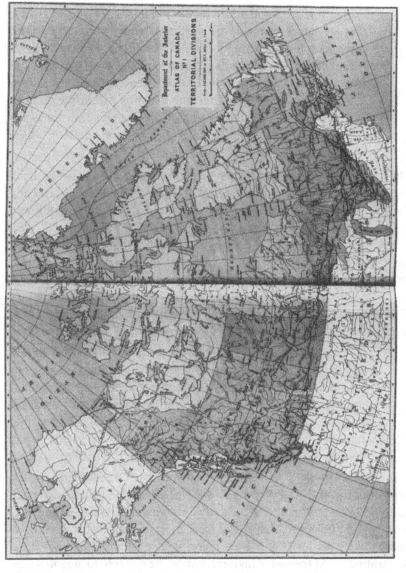

Fig. 3. Atlas of Canada, No. 1, 1906. *Source:* Public archives of Canada.

9

extending north of the St Lawrence up to the most hyperborean extremities – up to the North Pole'.[12]

Senator Poirier went on to refer to the visit of Captain Bernier to the Arctic Club in New York where:

> . . . it was proposed and agreed – and this is not a novel affair – that in future partition of northern lands, a country whose possession today goes up to the Arctic regions, will have a right, or should have a right, or has a right to all the lands that are to be found in the waters between a line extending from its eastern extremity north, and another line extending from the western extremity north. *All the lands between the two lines up to the North Pole should belong and do belong to the country whose territory abuts up there.*[13]

The Senator continued and systematized the sector theory just outlined by allowing a sector to Norway, Sweden, Russia, the United States and Canada. It will be noted that, without explanation, no sector was allowed to Denmark.

After Poirier's speech, the Honourable Sir Richard Cartwright, the representative of the Government in the Senate, intervened to say that, although 'Canada has a very reasonably good ground to regard Hudson Bay as a *mare clausum* and as belonging to it, that everything there may be considered as pertaining thereto', he went on to make the following reservation to a claim as far as the North Pole:

> Touching the other point my Honourable friend has raised, whether we, or whether any other nation is entitled to extend its territory to the North Pole, I would like to reserve my opinion. I am not aware that there have been any original discoverers as yet who can assert a claim to the North Pole, and I do not know that it would be of any great practical advantage to us, or to any country, to assert jurisdiction quite as far north as that.[14]

Senator Cartwright went on to point out that Canada had sent out expeditions in the Arctic and had established certain posts, had levied customs duties and exercised authority over whaling vessels. He added: 'I would think, however, that it was scarcely expedient for us, bearing in mind that a conference is now going on, to enter into any formal declaration'.[15] At the end of his intervention, he referred to 'negotiations' which were going on and said that 'it may not be the part of policy to formally proclaim any special limitation or attempt to make any delimitation of our rights there'.[16] He concluded by saying to Senator Poirier that he would do well not to press his motion and he himself moved that the debate be adjourned for the present, and it was so adjourned.

1.6. Summary

The use of meridians as a convenient method of delimiting a territorial claim goes back to the fifteenth century at least. In the Arctic, Canada seems to have been the first country to have so delimited its claim to lands and islands north of the continent. It first used the 141st meridian of longitude in its address of 1878 to Britain and then used both the 141st and the 60th meridians of longitude in 1897 to describe the boundaries of Franklin District. The description itself does not mention the 60th meridian, although the latter is used as a boundary on the map to which the Order in Council refers. The sector lines, however, do not extend to the North Pole but only to $78\frac{1}{2}°$ latitude in the west and 85° in the east. The full sector was indicated for the first time in 1904, on a map published by the Department of the Interior showing the explorations in Northern Canada. In these circumstances, Senator Poirier's motion of 1907 claiming 'all the lands between the two lines up to the North Pole' is not surprising. It should be noted that Canada's claim thus far has always been limited to lands and islands within the sector. As for the British transfer of 1880, it does not mention the sector lines and confines itself to territories, possessions and islands.

2

Boundary treaties as a legal basis for the sector theory

The 1825 Treaty between Great Britain and Russia, and the 1867 Treaty between the United States and Russia, have sometimes been invoked as a legal basis for the sector theory. Numerous writers and commentators have disagreed with such an interpretation of those treaties, and the matter is still the subject of controversy. This chapter examines the view of the main proponents of the boundary treaty basis and attempts a thorough analysis of the relevant provisions of those treaties.

2.1. Main proponents of boundary treaties
The main proponent of the sector theory is the Soviet writer, W. L. Lakhtine. He is also the one who tried to find a treaty basis for a full systematization of the theory in the Arctic. Before him, an American, David Hunter Miller, had given partial support to the theory and to the use of the boundary treaties for this purpose.

David Hunter Miller
Miller wrote in 1925, shortly after the Canadian Minister of the Interior, Charles Stewart, had used the sector theory to claim sovereignty right up to the Pole. Miller quoted from Article III of the 1825 Treaty which provides for the 141st degree of longitude to form the limit between the respective possessions of Russia and Great Britain in the following terms:

> . . . la même ligne méridienne du 141ème degré formera, dans son prolongement jusqu'à la Mer Glaciale, la limite entre les Possessions Russes et Britanniques sur le Continent de l'Amérique Nord-Ouest.[17]

Miller then pointed out that, in the 1867 Treaty which ceded Alaska to the United States, the description of the eastern limit was taken from the 1825 Treaty and translated as 'the said meridian line of the 141st degree, in its prolongation as far as the frozen ocean'.[18] He contended that 'the expression "as far as the Frozen Ocean" is vague enough (taking into account the previous treaty of 1825) to make it at least *arguable that the line runs as far as the 141st meridian itself runs,* and that means to the North Pole (for the continuation of that line beyond the Pole, it is not the 141st but the 39th meridian)'.[19]

Miller points out that the western limit prescribed in the 1867 Treaty follows a meridian which passes between certain named islands (at approximately the 169th degree longitude) and 'proceeds due north, without limitation, into the same Frozen Ocean'. He comments that 'these words "without limitation" are pretty strong words. They come very near to fixing the territorial rights of Russia and the United States, so far as those two countries could then fix them, up to the Pole'.[20]

Immediately following the above comment, Miller draws the following general conclusion:

> So I think we may say that the Canadian theory is, in part at least, based on the history of these treaties. It comes to this: the areas round the North Pole, whatever they may be, form three or four great cone-shaped sectors – the Canadian sector from 60° west to 141° west; the American sector from 141° west to 169° west; and the great Russian sector running from 169° west to some undefined line in the neighborhood of 30° or 40° east longitude. The remainder of the circle from say 40° east to 60° west, would, so far as this theory goes, be unassigned, but, very fittingly, that remainder seems to contain no land at all north of Spitsbergen and Greenland. Possibly a few islands close to the north Greenland coast are exceptions to this statement. *Whatever may be said by way of argument against this Canadian theory, it is certainly a highly convenient one.* All unknown territory in the Arctic is appropriated by three great Powers and divided among them on the basis of the more southerly status quo. Certainly if these three Powers are satisfied with such a partition, the rest of the world will have to be.[21]

In brief, the opinion of Miller is that, due to the imprecision of the terms used in the 1825 and 1867 Treaties, it is possible to maintain that the territorial rights of Russia, Canada and the United States were fixed right up to the Pole.

W. L. Lakhtine

In 1929, the Soviet jurist, W. L. Lakhtine, attempted to find a treaty basis for six Arctic sectors (see Figure 4).[22] He began by affirming that 'according to the now prevailing international law of the Arctic, these regions lie within the coastline of each of the polar states and the meridian limits running up to the North Pole'.[23]

The American sector is justified as follows:

> The boundaries the least open to dispute are the limits of the North American sector, which were established by the treaties of 1825 and 1867. From the Canadian side, the boundary runs along the meridian line of 141° west longitude, having been established by the Anglo–Russian Convention of 1825. From the side of the U.S.S.R., it runs approximately along the meridian line of 169° west longitude, fixed by the Russian–American Convention of 1867, and more precisely by the decree of the Presidium of the Central Executive Committee of the U.S.S.R. of 1926; that is to say, along the meridian line of 168°49′32″ west longitude.
> There is no ground for objection to this new line on the part of the United States, for it enlarges the American sector.[24]

The obvious criticism which must be addressed to the above passage is that the 1867 Convention, on which it relies to fix the boundary to follow the meridian line of 168°49′32″ west longitude, does not include any mention of that meridian or of any other specific meridian for that matter. As pointed out by the editor of the American Journal of International Law, the error of Lakhtine is to have assumed that the 1867 Convention had established the 169th west longitude as constituting the boundary.[25] If the 1867 Convention had in fact fixed the boundary at the 169th meridian, the Soviet Decree of 1926 could not possibly have changed it unilaterally. However, what is much more questionable than the exact location of the boundary is the presumption by Lakhtine that the meridian in question was meant to be a boundary right up to the North Pole. He advances no argument whatever to support that contention.

Lakhtine describes the sector of the Soviet Union in the following terms:

> On the east its boundary passes along the same meridian line of 168°49′32″ west longitude, and on the west it borders upon the Norwegian–Finnish sector, the boundary being the meridian line of 32°4′35″ east longitude according to the decree of the Presidium of the Central Executive Committee of the U.S.S.R. of 1926. This boundary line curves into the sector of the U.S.S.R. at

Fig. 4. Sector lines proposed by W. L. Lakhtine, 1930. *Source*: Prepared for the author by John E. Cooper, Technical Advisor to Dept of External Affairs, Ottawa.

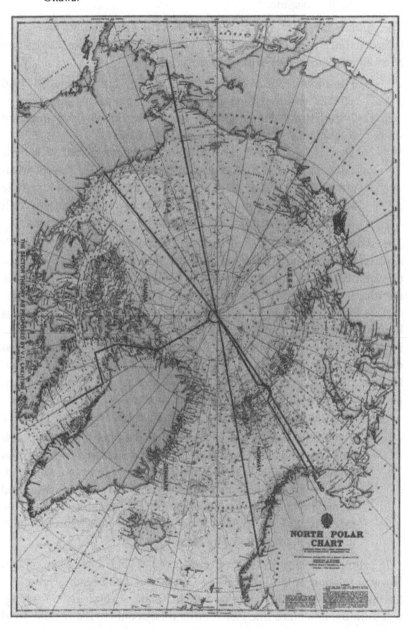

about 35° east longitude in the region of the Spitsbergen Archipelago belonging to Norway. This boundary also is not open to dispute.[26]

The author does not say what is the basis for fixing the boundary line between the Soviet Union and Norway at 32°4'25". As for the Finnish and Norwegian sectors, they are described as follows:

The Finnish sector lying within the meridian lines of 31° and 32°4'35" east longitude approximately, and having on a level of 70° north latitude a base width of only one degree, is so small that it cannot be of much importance to Finland.

The eastern boundary of the Norwegian sector, fixed by the Paris Convention of 1920, has already been mentioned, and the western boundary passes along the meridian line of 10° east longitude approximately. Denmark can raise no objection to this line, as it is contiguous to that of Norway and fully satisfies her claims.[27]

The Paris Convention of February 9, 1920 does recognize Norway's sovereignty over the Archipelago of Svalbard and the islands situated between 10° and 35° longitude, but it is difficult to understand how these meridians could be used as a conventional basis over the whole of the sector. Indeed, the Paris Convention specifies that the area envisaged is located between 74° and 81° latitude north.[28] In other words, far from implying that Norway could claim a sector between the 10th and the 31st degrees of longitude up to the Pole, the Treaty makes it quite clear that Norway's sovereignty is limited to the islands between those meridians and that the northern limit of those islands is the 81st degree of latitude.

As for the Danish sector, it is described as follows:

The Danish sector lies between the meridian lines of 10° east longitude and of 60° west longitude approximately, with a certain curve, advantageous to Denmark, along the coast of Greenland. It is fixed by the Convention of 1924 between Norway and Denmark and has been recognized by England. So that the extent of this sector also is acknowledged by the States that have jurisdiction over the contiguous sectors.[29]

It is difficult to see how the Convention of 1924 between Denmark and Norway relating to the area east of Greenland could be interpreted as having established a sector within which Danish sovereignty was recognized. In the first place, the Convention relates only to the regulation of fishing and hunting between the parties and does not touch the question of sovereignty. As pointed out by the Permanent Court of International Justice when it awarded sovereignty over East Greenland to Denmark, '[t]he question of sovereignty . . . was completely left out of the Convention of 19 July 1924'.[30] In the second place, no mention whatever is made of meridians of longitude in the Treaty. Only Article I of the treaty speaks of meridians and both of them are parallels of latitude. The meridian of 10° longitude cannot be said to constitute a boundary by virtue of the Convention of 1924.

As for the Canadian sector, Lakhtine describes it as follows:

> Lastly the Canadian sector lies between the western Danish boundary of the sector, i.e., approximately between the meridian line of 60° west longitude and the eastern boundary of the American sector, i.e., the meridian line of 141° west longitude. These limits are based on the Convention of 1825 and the agreement with Denmark.[31]

It is impossible to understand what agreement Lakhtine could have had in mind, laying down a boundary line between Denmark and Canada along the 60th meridian of longitude. It will be recalled that in 1920 Denmark allegedly subscribed to a view which had been expressed in a report by Rasmussen that Ellesmere island was no man's land, which resulted in a strong protest by Canada through Great Britain. However, no agreement was ever concluded between the two countries, and there was no mention of meridians in the correspondence.

We will examine next the assertion that the western boundary of the Canadian sector follows the meridian of 141° west longitude and that it finds its conventional basis in the 1825 Treaty between Russia and Great Britain.

2.2.   The 1825 Boundary Treaty

The problem of interpretation facing us concerns the description of the line of demarcation found in Article III of the 1825 Treaty.[32] More specifically, the question is whether or not the intention of the Parties was to use the 141st degree of longitude as a boundary extending across the Arctic Ocean right up to the North Pole. The problem of interpretation arises out of the use of the following expressions: 'jusqu'à', 'Mer Glaciale', 'ligne de démarcation' and 'possessions'.

According to the well-known general rule of interpretation, consistently followed by the International Court and incorporated in the Vienna Convention on Treaty Law in 1969, the terms of the Treaty must be given their ordinary meaning in their context and in the light of the Treaty's object and purpose. No special meaning is to be given to a term unless it is established that the Parties intended such a meaning. The context, for the purpose of interpretation, includes both the text of the

Treaty itself as well as the preamble and annexes. Account may also be taken of any subsequent practice in the application of the Treaty, as well as of any subsequent agreement between the Parties regarding its interpretation or application.

If the ordinary meaning leads to an ambiguity or to a manifestly absurd or unreasonable result, recourse may be had to the preparatory works of the treaty and to the circumstances of its conclusion as supplementary means of interpretation. Resort to these may also be made in order simply to confirm the ordinary meaning.

Applying the rules of interpretation just reviewed, one notes, through a reading of the preamble of the Convention, that the intention of the parties was to settle 'different points connected with the commerce, navigation and fisheries of their subjects of the Pacific Ocean, as well as the limits of their respective possessions on the northwest coast of America'.[33] The rights of commerce, navigation and fishing are settled in Articles I and II of the Treaty and the limits of the respective possessions are defined in Article III.

The introductory clause of Article III reads:

> La *ligne de démarcation* entre les Possessions des Hautes Parties Contractantes sur la Côte du Continent et les Iles de l'Amérique Nord-Ouest, sera tracée ainsi qu'il suit:[34]

The line then described commences at the southernmost point of Prince of Wales Island, determined in the 1903 Alaskan Boundary award[35] as being Cape Muzon, continues in a general northerly direction along the Portland Channel to the 56th degree of latitude, and follows the summit of the mountains until it reaches the 141st meridian of longitude. The line continues from that point pursuant to the following description:

> . . . la même ligne méridienne du 141ème degré formera, dans son prolongement *jusqu'a* la *Mer Glaciale,* la limite entre les *Possessions* Russes et Britanniques sur le Continent de l'Amérique Nord-Ouest.[36]

Let us now take one by one the four key terms or expressions and try to interpret them in light of the rules of interpretation just recalled.

Meaning of 'jusque'

With respect to the expression 'jusque', the question arises whether or not it is inclusive or exclusive of the object to which it relates. The ordinary meaning of the preposition 'jusque', as defined in the Universal French Dictionary of the 19th Century, is 'le terme, la limite que l'on atteint, sans la dépasser: aller jusqu'à Bordeaux. Dieu a dit à la

mer: Tu viendras jusque-là et tu n'iras plus loin'.[37] All of the examples
given by the dictionary support the meaning that the preposition *'jusqu'à'*
indicates either a limit in time or in space which is reached but not exceeded.

The question might nevertheless be asked as to whether there might not
be a difference between reaching a place without entering it, and reaching
a topographical feature without crossing it. This question did arise in the
case of the *Monastery of St Naoum* in 1924. In its Advisory Opinion, the
Permanent Court of International Justice had to decide whether the
monastery in question was included within the boundaries of Albania.
The expression 'jusqu'à' was contained in the description of the boundaries of the newly created State of Albania, as drafted by a delimitation
Commission and incorporated in a Protocol. The relevant part of the
description read as follows:

> La région côtière jusqu'à Phtelia, y compris l'île de Saseno, la
> région au nord de la ligne grecque, ainsi que l'ancien caza de
> Koritza avec la rive ouest et sud du lac Orhrida s'étendant du
> village de Lin jusqu'au monastère de Svetnaoum feront partie de
> l'Albanie.[38]

The Court decided that the monastery in question was intended to be
included within the boundaries of Albania, but did not do so by making a
choice between the two possible interpretations of the preposition
'jusqu'à'. Indeed, the Court specifically stated that it was not making any
such choice.[39] However, interpreting the term in its immediate context
and considering that the Parties had definitely intended to include
another place, namely Phtelia, which was mentioned in exactly the same
way after the same preposition, it concluded that the latter ought to be
given the same interpretation. Consequently, the judgment of the Court,
by itself, is not very helpful.

The oral argument, however, presented by Professor Gilbert Gidel on
behalf of Albania, is most enlightening. Approving the distinction made
by one of the expert witnesses, Professor Gidel distinguished between the
case where the name of the place following the preposition 'jusquà' is a
town or a monastery in which it is possible and normal to enter, on the one
hand, and the name of a place or location which is a topographical feature
and serves as a means of separation, such as the summit of a mountain or
a river, on the other. In the latter case, the preposition 'jusque' is given
the ordinary meaning of 'up to but no further' than the location specifically
mentioned. The relevant part of Professor Gidel's oral pleading reads as
follows:

> Voilà la vérité. Il ne faut pas dire: 'jusque' suivi d'un lieu veut

toujours dire 'à l'exclusion du lieu' ainsi désigné. Il faut distinguer quel est ce nom de lieu. S'il s'agit d'un élément topographique de séparation, comme le dit parfaitement le général Tellini, s'il s'agit d'une crête ou d'une rivière, 'jusque' veut dire: *'en s'arrêt-ant à cet élément topographique de séparation' sans le dépasser, sans y pénétrer,* jusqu'au contact de cet élément topographique. Mais si 'jusque' est accolé à un nom de lieu qui ne soit pas un élément topographique de séparation, 'jusque' veut dire 'et y compris'. C'est le sens qu'a 'jusque' dans l'exemple que je prenais de cet ami qui, ayant voulu m'accompagner jusqu'à Paris, n'avait pas été empêché par les exigences de la langue française de pénétrer dans Paris avec moi.[40]

Applying the ordinary meaning of the term 'jusqu'à' to the present situation, it would seem that *the end of the boundary line along the 141st degree of longitude was intended to be at its point of intersection with 'la Mer Glaciale',* the topographical element chosen by the parties. This interpretation is also in accordance with the meaning obviously intended for the other two turning points in the same description where the preposition 'jusqu'à' was used; more specifically, as far as the 56th degree of north latitude and as far as the point of intersection of the 141st degree of west longitude.

Meaning of 'Mer Glaciale'

The second expression causing some difficulty is 'Mer Glaciale'. It was translated into English as 'Frozen Ocean' which is, as properly noted by Miller, rather vague. If taken literally, it could even be interpreted as meaning that the boundary line along the 141st degree was intended to run as far as that part of the Arctic Ocean which is permanently frozen or covered with ice. This would have meant somewhere around 74° north latitude. However, there are at least two reasons militating against such an interpretation. Firstly, only the French text is authentic, and the meaning of the adjective 'glaciale' is not properly rendered by 'frozen'. Secondly, the expression 'Mer Glaciale' was synonymous with the expression 'Océan Glacial' which was used at the time of the conclusion of the treaty to indicate either the Arctic Ocean or the Antarctic Ocean, depending on the context. The Universal French Dictionary of the 19th century defines the expression 'Mer Glaciale' as follows:

Océan Glacial ou *mer Glaciale:* nom que l'on donne à la mer, couverte de glace en grande partie, qui s'étend du pôle nord au cercle polaire arctique, ainsi qu'à la mer supposée qui occupe une position semblable au cercle polaire antarctique: *océan Glacial arctique.*[41]

It should be reasonably clear from the above that the intention of the Parties was to use the Arctic Ocean as a topographical feature indicating the limit of the demarcation line between their respective positions. However, if it is still argued that it might have been the intention of the Parties to use the 141st meridian of longitude as a boundary across the Arctic Ocean as far as the North Pole, it must be recalled that, even in 1825, this would have been against the principle of the freedom of the high seas, already established and subscribed to by both Great Britain and Russia. The extent of the territorial sea and the exact nature of the coastal state jurisdiction over waters adjacent to its coast, however, had not been established.

The issue of the status of coastal waters had been raised by the 1821 Russian ukase (proclamation) which had claimed fishing and sealing jurisdiction up to 100 miles offshore. One of the purposes of the 1825 treaty, as previously indicated, was to settle the questions of commerce, navigation and fishing rights between the parties in the Pacific Ocean, because of the objections to the ukase made by Great Britain and the United States. Consequently, it would not be interpreting the treaty in the light of its stated purpose if the demarcation line were said to run as far as the North Pole across the Arctic Ocean.

Meaning of 'ligne de démarcation'

The third expression which should be examined is 'ligne de démarcation'. That expression was and is still used in relation to the establishment of a boundary and indicates the actual course which the boundary should follow. It was in that sense, for instance, that the expression was used in the 1815 Paris Peace Treaty.[42] The term 'boundary' is associated with land and delimits a special zone over which a State has complete jurisdiction.[43]

That the demarcation line described in the 1825 Treaty was meant to apply to land only is evident from the introductory clause of Article III.[44] That clause specifies that the line of demarcation is to be 'upon the *Coast* of the Continent, and the *Islands* of America to the Northwest'.[45]

Meaning of 'possessions'

A fourth expression which might lend itself to more than one interpretation is 'possessions'. This term is used in the title of the Convention, in the preamble and throughout the Treaty, in particular in

Article III. Was the expression meant to indicate land possessions only or was it intended to include part of the Arctic Ocean? According to a well-known dictionary of international law terminology, the term 'possessions' is normally used to refer to the overseas territories or colonies of a State.[46] For instance, in the Consular Convention between France and Russia in 1874, it is specified that each contracting party may appoint consuls in the ports of the territory of the other party, including overseas possessions and colonies.[47] We find in the same dictionary that the term 'territory' normally refers to all land areas over which a state exercises sovereignty, including territorial waters.[48] In other words, the ordinary meaning of the term 'possessions' is limited to land territory, internal or inland waters and perhaps also territorial waters.

This ordinary meaning seems to have been intended by the Parties to the 1825 Treaty since they specified in Article III that the 141st meridian was to be the limit 'entre les *Possessions* Russes et Britanniques *sur le Continent';* in other words, the subject matter of the treaty was 'possessions on the Continent'. It might also have included territorial waters, but it is not at all certain that the concept was sufficiently well established in 1825 to be so included. An additional element of proof that the term 'possessions' was meant to be given its ordinary meaning is found in the introductory clause of Article III where the line of demarcation between the possessions of the parties is described as being 'upon the Coast of the Continent and the Islands of America to the North-West'.[49]

The above analysis of the 1825 Treaty leads to the conclusion that the Treaty could hardly serve as a legal basis for the sector theory as contemplated by Mr Miller and Mr Lakhtine. However, it has been argued that the 1867 Treaty between the United States and Russia could perhaps be invoked for such a purpose.

2.3.   The 1867 Boundary Treaty

The 1867 Treaty, which ceded Alaska to the United States, described the eastern limit along the 141st degree of longitude by incorporating the description found in the 1825 Treaty. The western limit is described as follows:

> The western limit within which the *territories and dominion* conveyed, are contained, passes through a point in Behring's straits on the parallel of sixty-five degrees 30 minutes north latitude, at its intersection by the *meridian which passes midway between the islands* of Krusenstern, or Ignalook, and the island of Ratmanoff, or Noonarbook, and *proceeds due north, without limitation, into the same Frozen Ocean.*[50]

In order to assess the possible validity of the argument which uses this provision as a treaty basis for the sector theory, it is necessary to examine at least four terms or expressions: 'territories and dominions'; 'meridian which passes midway . . .'; 'without limitation'; and 'Frozen Ocean'.

Meaning of 'territories and dominion'

The ordinary meaning of 'territory', as indicated earlier with respect to the 1825 Treaty, is land territory. And when this term is associated with dominion, as it is in the 1867 Treaty, that ordinary meaning would seem to be reinforced. In that year, when Canada was granted dominion status, it meant that Canada had complete internal autonomy and full control over its territory. There is no indication in the 1867 Treaty that the parties intended to give to this expression anything but the ordinary meaning. In other words, when Russia agreed to cede 'all the territory and dominion now possessed by His Said Majesty on the continent of America and in the adjacent islands, the same being contained within the geographical limits herein set forth',[51] it was transferring title to the United States of the 'lands and islands' over which she had full control and sovereignty. The cession could hardly have included anything beyond the ordinary territorial waters.

This interpretation is also supported by the award of the tribunal in the *Bering Sea Seal Fishery Arbitration* in 1893. In that case, the United States claimed that, on the basis of the 1867 Treaty, it had exclusive jurisdiction to hunt for seals in the Bering Sea beyond the three-mile limit. The Tribunal held that Russia itself had no exclusive fishing rights beyond the ordinary limit of territorial waters and, therefore, could not have transferred any such rights to the United States in 1867.[52]

Meaning of 'meridian which passes midway between the islands'

The second expression which should be examined is 'the meridian which passes midway between the islands of Krusenstern, or Ignalook, and the island of Ratmanoff, or Noonarbook'. The least that can be said about this meridian, as a possible boundary, is that it is not specifically identified. It could be the 169th degree of longitude or the 168°49'30" as mentioned in the 1926 Decree of the Soviet Union and repeated by Lakhtine. It could also be the meridian of 168°52'22.587" of longitude, as adopted by a group of U.S. cartographic experts.[53] Considering this uncertainty, it is quite unlikely that the parties could have intended the meridian in question to constitute a boundary equivalent to a land boundary.

Meaning of 'without limitation'

The next expression, 'without limitation', causes more difficulty and constitutes the main basis for the proponents of the sector theory to argue that the meridian was meant to be a boundary right up to the Pole. It may be contended that the intention of the parties was definitely to prolong this boundary right up to the Pole, since those words did not appear in the 1825 Treaty with respect to the 141st meridian.

In trying to determine the real intention of the parties, one must look at both the English and French versions of the 1867 Treaty, since both are supposed to be authentic texts. The English version uses the expression 'without limitation into the same *Frozen* Ocean' whereas the French version reads 'sans limitation, vers le nord, jusqu'à ce qu'elle se perde dans la Mer Glaciale'. Regardless of which text one takes, there seems to be considerable ambiguity in that a boundary line, if the meridian in question is supposed to serve as one, must have a limit. The French text is even more confusing in that the expression 'sans limitation' leads to an *unreasonable result* when applied to a boundary, and the expression 'jusqu'à ce qu'elle se perde dans la Mer Glaciale' is *ambiguous* in that it is difficult, indeed impossible, to determine the point at which the boundary may be said to lose itself in the ocean. In both cases, resort must be had to the full context of the treaty and also, preferably, the circumstances surrounding its conclusion.

The rest of the context of the treaty makes it quite clear that the object of the cession or transfer was territorial. Indeed, every single article of the treaty (except Article VII which pertains only to ratification) speaks of 'territories and dominion', 'territory and dominion', 'territory or dominion' and 'territory'. One article enumerates the following: 'territory, dominion, property, dependencies and appurtenances' as constituting the object of the cession in question (Article IV). In other words, *there is no indication whatever throughout the treaty that the parties contemplated any part of the Arctic Ocean.*

The above contextual interpretation is supported by an examination of the circumstances surrounding the conclusion of the treaty. It should be recalled that, in 1867, the parties to the treaty believed in the existence of a continent which they thought was situated somewhere north of Alaska and which they were trying to discover. Indeed, the explorers of the time were still in search of a northern continent commonly called 'Arctic Continent'. This was particularly the case for the Russian and American explorers. In 1820, the Russian Imperial Government had sent the Wrangel expedition in search of the Arctic continent. Then, when Herald and Wrangel islands were discovered by Captain Kellett, in 1849, it was thought that they were part of the famous continent. The explorer, V. Stefansson, tells us that this was still the case in 1879:

> The hypothetical continent was still in the minds of scientists when Lieutenant De Long was fitted out by the New York 'Herald' in 1879. He steered the *Jeannette* boldly northward from the Pacific into the ice beyond Bering Straits, thinking that he could not drift far, for the 'continent' would bar the way. But, fast in the pack, he did drift far – right across the theoretical continent and beyond what now proved to be Kellett's Island rather than Kellett's Land.[54]

Considering that the contracting parties of the 1867 Treaty were still in search of a continent somewhere north in the Arctic Ocean, it is most unlikely that they would have intended to extend their boundaries to or across that continent. Even if they did, they would have had land and not sea boundaries in mind.

Meaning of 'Frozen Ocean'

Finally, it must be noted that the expression 'Frozen Ocean' is obviously borrowed from the translated version of the 1825 Treaty, and the comments made earlier in this Section apply fully here. Since the expression is meant to refer to the Arctic Ocean, it does not indicate an intention of the parties to run the boundary line as far as the permanently frozen edge of that ocean.

A final, and it would appear irresistible, argument against considering the 141st and 169th (approximately) degrees of longitude as sector boundary lines is found in the governing part of Article I of the treaty. This article specifies that what is being ceded to the United States by the Convention is the 'territory and dominion' and the 'adjacent islands' situated 'within the *geographical limits* herein set forth'.[55] Since the treaty then goes on to describe the 'eastern limit', as being the 141st degree and the 'western limit', as being the meridian which approximates the 169th degree, it appears quite obvious that *the meridians in question are merely used to establish the geographical area within which the land and islands forming the object of the transfer were located.*

The above interpretation is supported by the very words of formal transfer used by the Russian Commissioner on October 18, 1867: 'I transfer . . . all the territory and dominion now possessed on the continent of America and in the adjacent islands, according to . . .'[56]

With respect to the interpretation just given of the 1867 Convention, reference should be made to the 1955 study by a group of U.S. Cartographic experts entitled 'Coordinate Positions for the Plot of U.S.–Russia Convention of 1867'.[57] This document has been adopted as the standard description for the cartographic representation of the 1867 Convention and fixes the northernmost point of the line at 72° north. The explanation in the study reads as follows:

> It should be noted that the original Convention language stated that the line 'proceeds due north, without limitation, into the same Frozen Ocean'. *Since the United States does not support so-called 'sector claims' in the polar regions, the northernmost point for the representation of the Convention line was agreed to be 72°00′N.* Furthermore, in keeping with the policy that the line does not constitute a boundary, the standard symbol for the representation of an international boundary should never be used. Furthermore, labeling of the line as 'U.S. – Russia Convention of 1867' is recommended.[58]

The above view that the convention line does not represent a boundary, but rather simply a line of allocation of land, is consonant with that expressed by the well-known geographer S. W. Boggs. He concluded that 'most lines in water areas which are defined in treaties are not boundaries between waters under the jurisdiction of the contracting parties, but a cartographic device to simplify description of the land areas involved . . .'[59] That judgment would certainly appear to be applicable here.

2.4.   Summary

The above analysis leads to the conclusion that *the 1825 and 1867 boundary treaties cannot serve as a legal basis for the sector theory,* at least not in the sense of being capable of giving title to sea areas. The meridians were used as convenient geographical devices with which Russia and Great Britain delimited their possessions along the coast and the islands in 1825, and within which the dominion or sovereignty over the territory was ceded by Russia to the United States in 1867. In both instances, the subject matter was land only, and not land and sea, except for inland waters and the territorial sea.

This interpretation would appear to be one given by Canada itself when it adopted an Act to provide for the government of the Yukon District in 1898. The Act incorporates in its Schedule a Proclamation, establishing the boundaries of the Yukon Judicial District, which describes the boundaries as 'beginning at the intersection of the 141st meridian of west longitude from Greenwich with a point on the coast of the Arctic sea, which is approximate north latitude, 69°39′′, and named on the Admiralty charts "Demarcation Point" ' and ending by 'following the winding of the Arctic Coast (termination of the mainland of the Continent), including Herschel Island, and all other islands which may be situated within three (3) geographical miles, to the place of beginning'.[60]

It will be noted that the boundaries do not extend beyond the coast, except for islands situated within three miles. Such islands within the territorial waters had already been held in international law to come automatically under the sovereignty of the coastal State by the simple fact of their close proximity or contiguity.[61]

3

Contiguity as a basis for the sector theory

It is generally recognized that the doctrine of contiguity forms the basis of the sector theory. Even those who invoke the boundary treaties of 1825 and 1867 look to that concept to reinforce their case. This section attempts to determine the extent to which States and international tribunals have relied on contiguity. This is followed by an appraisal of this concept as a principle of international law capable of generating title or sovereignty.

3.1.   Contiguity in State practice

The notion of contiguity appeared in international law when principles for the acquisition of sovereignty over new territories were being developed. Acquisition through the act of discovery, accompanied by a formal taking of possession, having led to abuses, the Roman Law principle for the acquisition of property was introduced. It required the fulfilment of two conditions: the open intent to occupy (animus) and the actual occupation of a well-defined property (corpus). The second condition being difficult or perhaps impossible to fulfil in many instances, the doctrine of contiguity or geographic proximity was developed.

Under this doctrine, the effective occupation of part of a region or territory gave title to the whole of the unoccupied region or territory proximate enough to be considered as a single geographic unit with the occupied portion. The same doctrine, with the occasional nuance, has been presented under different names, in particular the following: proximity, propinquity, hinterland, adjacency, continuity, geographic unity and region of attraction.

The first use by a State of the doctrine of contiguity seems to have been made in 1844 during the *Oregon controversy* between the United States and Great Britain. The contested Oregon territory on the west coast of the American continent was situated west of the Rocky Mountains between latitudes 42° and 54°40′ north. In support of the American position, Secretary of State Calhoun wrote to the British Minister in the following terms:

> That continuity furnishes a just foundation for a claim of terri-
> tory, in connection with those of discovery and occupation,
> would seem unquestionable. It is admitted by all, that neither of
> them is limited by the precise spot discovered or occupied . . .
> but how far, as an abstract question, is a matter of uncertainty. It
> is subject, in each case, to be influenced by a variety of consider-
> ations.[62]

The controversy was finally settled by a treaty in 1846 which fixed the boundary at the 49th degree of latitude. Is it possible to deduce from the settlement some acceptance of the doctrine of contiguity by the Parties? A study of the relevant diplomatic correspondence leads to a negative answer for at least two reasons.[63] Firstly, the United States relied on some five other bases: the possibility that the 1803 cession of Louisiana might have included the territory west of the Rockies; the Spanish rights transferred to the United States in 1819; the discovery and exploration of the Columbia river in 1792 by Captain Gray; the explorations by Lewis and Clarke in 1804–06; and the establishment of fur trade posts in the area. The second reason is that Secretary James Buchanan, Calhoun's successor, who continued the negotiations with British Minister Pakenham to the conclusion of the treaty, abandoned the argument based on contiguity.

The position of the United States on the contiguity doctrine was made clear in two subsequent controversies: one in 1852 relating to the Labos Islands and the other, in 1873, relating to the Island of Navassa. In the dispute with Peru over the *Labos Islands* situated about 30 miles from the Peruvian coast, Secretary of State Webster refuted the contiguity argu-
ment presented by Peru. He wrote that unequivocal acts of sovereignty were necessary to acquire title to the islands because they were 'so far from any continental possession of Peru as not to belong to that country by the law of proximity or adjacent position'.[64] Mr Webster went so far as to deny that the proximity of the islands even gave rise to a prima-facie claim to them because their distance from shore was 'five or six times greater than the three marine miles extend.'[65]

The United States maintained its negative position on contiguity in its dispute with Haiti relating to the *Island of Navassa*. Secretary of State Fish wrote to the Haitian Minister that there had been no acts of jurisdiction by Haiti and concluded that this 'must . . . appear to any reasonable mind fatal to that claim, nor can this absence be supplied by the facts of contiguity'.[66]

During the nineteenth century, various European powers invoked the concept of contiguity, claiming that their occupation of part of the coast of the African continent gave them title to an indefinite area of the hinterland or interior country. In order to prevent abuses resulting from the application of this concept, the *General Act of Berlin of 1885* laid down two essential conditions for the acquisition of territorial sovereignty in Africa: (1) effective occupation of the territory claimed by the establishment of an authority sufficient to insure order and commercial freedom, and (2) notification to other interested Parties. Although the principle of effective occupation eventually became part of customary international law, it bound only the Parties to the treaty at the time and was applicable solely to the African continent. Contiguity continued to be invoked in other parts of the world.

In the Arctic, reliance on the concept seems to have started in 1907 when *Canada's* Senator Poirier tried to find a legal justification for a claim to all lands and islands within his proposed sector. He invoked 'the theory that the frontage of the seaboard carried with it the strip of land all the way across the continent'.[67] He was referring, of course, to the hinterland theory.

In its note of 1916, *Russia* indirectly invoked the concept of contiguity or continuity when it claimed the islands which it qualified as being 'close to the asiatic coast of the Empire and a northern extension of the Siberian continental platform (platforme continentale)'.[68] The 1924 note of the Soviet Union used the same expression, 'close to the asiatic coast', and the note accompanying the 1926 Decree referred to 'the part of the Arctic regime adjacent to the northern coasts of the Union'.[69]

On November 25, 1920, *Canada's* Under-Secretary of State for External Affairs, Joseph Pope, relied on the theory of contiguity when he stated in a memorandum that 'Our claim to the islands north of the mainland of Canada rests upon quite a different footing (different from any claim Canada could have made to Wrangel Island), by reason of their geographical position and contiguity'.[70] In 1924, the Minister of the Interior, Charles Stewart, invoked the same concept in his answer to a question as to whether other countries were making claims to the islands to the north of Canada. He replied: 'Of course my Honourable friend is aware that international law, in a vague sort of way, creates ownership of unclaimed lands within one hundred miles of any coast, even if possession has not been taken. At least there is a sort of unwritten law in that respect.'[71]

There is also evidence that some Antarctic sector claims are based, at least partly, on the concept of contiguity. In 1929, the British Minister in Oslo informed the Norwegian government that *'Great Britain* had unimpeachable rights to the whole of these sectors, including all land down to the South Pole, an extension of which was looked upon as the inseparable hinterland of the coastal territory in each sector'.[72] In an official declaration of September 1950, *Chile* stated that its 'claim is supported by logical geographic continuity and contiguity . . . in addition to actual permanent occupation . . .'.[73]

It must be noted, however, that in the *Eastern Greenland Case*, reviewed below, Norway took a very strong stand against contiguity and Denmark vigorously denied placing any reliance on the concept. There is no evidence that either of these countries has ever changed its position.

3.2. Contiguity in international decisions

Contiguity has been discussed in a number of international decisions dealing with land or islands and recently in some cases relating to the continental shelf. These two groups of cases are reviewed below.

Contiguity of lands and islands
Aves Island Case (1865)

In this dispute between Venezuela and The Netherlands, the latter relied partly on the proximity and former unity of the disputed island with Saba Island, over which it had uncontested sovereignty. The arbitrator rejected the argument of The Netherlands, since the disputed island was some 40 leagues away from Saba Island and 28 leagues from the sand bank which formerly united both islands.[74]

Island of Bulama Case (1870)

This case between Great Britain and Portugal concerned an uninhabited island very close to the west coast of Africa, between the river Jeba and the Rio Grande. Portugal was awarded the island by the arbitrator on three grounds: its discovery of the island, its settlement and sovereignty over the whole coast of the mainland opposite, and the fact that 'the Island of Bulama is adjacent to the mainland and so near to it that animals cross at low water'.[75]

In these circumstances, contiguity was accepted as a basis of title, considering the very close proximity and the additional fact that the only ground of Great Britain was 'an alleged cession by native chiefs in 1792, at which time the sovereignty of Portugal had been established over the mainland and over the Island of Bulama'.[76] In short, this was the case of an island in the territorial waters of the coastal State.

Monks Islands Case (1885)

This group of rocks, covered by a deposit of guano and lying some twenty miles off the coast of Venezuela, was briefly occupied by American citizens for the purpose of removing the guano. After their expulsion by Venezuela, which seized their equipment and materials, the United States brought this action for damages on behalf of its citizens. The question was whether the proximity of the islands to Venezuela was such as to make the temporary occupation a trespass and thus constitute a bar to the action.

The Claims Commission rejected the argument based on contiguity and, after specifying that this was not a sovereignty case, stated that 'in the absence of exceptional circumstances we can see no reason for distinguishing between the occupation of the Monks Islands and a similar unoccupied group in the mid-Pacific'.[77]

British Guyana Case (1904)

This case involved the determination of the boundary line between British Guyana and Brazil. Since the arbitrator could not decide with certainty whether the claim of Brazil or that of Great Britain was the stronger, he held that the contested territory 'should be divided in accordance with lines traced by nature' following the 'thalweg' of two rivers.[78] During his discussion of the principle of effective occupation, the arbitrator stated 'that the effective occupation of a part of a region, although it may be held to confer a right to the acquisition of the sovereignty of the whole of a region which constitutes a single organic whole, cannot confer a right to the acquisition of the whole of a region which, either owing to its size or to its physical configuration, cannot be deemed to be a single organic whole *de facto*'.[79]

The quoted passage calls for at least two comments: first, what may flow from contiguity in certain situations is not title or sovereignty but rather only a right to acquire it; second, the arbitrators did not find that this was a situation where contiguity could be applied.

Island of Palmas Case (1928)

This case between The Netherlands and the United States involved an inhabited and somewhat isolated island, situated approximately 45 miles from the Philippine Archipelago. The United States based its claim on three grounds: discovery by Spain in the 16th century; recognition of Spanish sovereignty by The Netherlands in signing the Treaty of Munster in 1648, and contiguity. The arbitrator, Max Huber, awarded the island to The Netherlands on the basis of the display of acts or manifestations of sovereignty judged to be sufficient in the circumstances.

In rendering his decision, the arbitrator dealt quite fully with the concept of contiguity. In the first place, the arbitrator held that contiguity cannot give title to an island outside territorial waters.

> Although States have in certain circumstances maintained that islands relatively close to their shores belonged to them in virtue of their geographical situation, it is impossible to show the existence of a rule of positive international law to the effect that islands situated outside territorial waters should belong to a State from the mere fact that its territory forms the terra firma (nearest continent or island of considerable size). Not only would it seem that there are no precedents sufficiently frequent and sufficiently precise in their bearing to establish such a rule of international law, but the alleged principle itself is by its very nature so uncertain and contested that even Governments of the same State have on different occasions maintained contradictory opinions as to its soundness.[80]

The arbitrator went so far as to say that, even as a rule creating a presumption in favour of a particular State, it would be in conflict with the necessity of displaying activities of a State.[81] He added: 'Nor is this principle of contiguity admissible as a legal method of deciding questions of territorial sovereignty; for it is wholly lacking in precision and would in its application lead to arbitrary results.'[82]

The arbitrator did admit, however, that contiguity could have a limited application at the time of taking possession of a group of islands. The relevant passage reads:

> As regards groups of islands, it is possible that a group may under certain circumstances be regarded in law as a unit, and that the fate of the principal part may involve the rest. Here, however, we must distinguish between, on the one hand, the act of first taking possession, which can hardly extend to every portion of territory, and, on the other hand, the display of sovereignty as a continuous and prolonged manifestation which must make itself felt through the whole territory.[83]

It will be noted that the arbitrator refused to extend the role of contiguity to the actual maintenance or continuous exercise of sovereignty.

Toward the end of his award, the arbitrator summed up his view on contiguity as a possible basis of title by saying: 'The title of contiguity, understood as a basis of territorial sovereignty, has no foundation in international law.'[84]

Eastern Greenland Case (1933)

In 1933, the contiguity theory was discussed at some length during the oral arguments in the *Eastern Greenland Case* between Denmark and Norway. Norway alleged that Denmark was relying on the doctrine in order to avoid showing effective occupation of the eastern part of Greenland. Denmark, however, vigorously denied that it was placing any such reliance on contiguity. In its 'Réplique', counsel for the Danish Government stated in particular: '. . . le Gouvernement danois n'a nulle part invoqué la théorie de la contiguité pour fonder son droit sur le Groenland tout entier'.[85]

In spite of the above denials, Professor Gidel, in his oral pleading for Norway, continued to contend that Denmark's arguments were strongly influenced by the doctrine of contiguity.[86] Professor Gidel's condemnation of the contiguity doctrine begins in the following terms:

> Comment d'ailleurs le Danemark aurait-il pu se réclamer nettement d'une pareille notion universellement condamnée?
>
> Le Contre-Mémoire norvégien contient la liste imposante des auteurs qui se sont prononcés contre la théorie de la contiguité.
>
> Il contient l'indication de nombreux cas de pratique internationale d'où se dégage la même condamnation.
>
> Enfin, comment prétendre se prévaloir d'un titre auquel sont consacrées dans la sentence Palmas ces deux lignes péremptoires: 'Le titre de la contiguité, entendu comme une base de souveraineté territoriale, n'a pas de fondement en droit international?'[87]

Regardless of whether or not Denmark was in fact placing some reliance on the contiguity doctrine, what is important is that both Norway and Denmark denounced it as a possible legal basis of territorial sovereignty.

The Court's judgment in favour of Denmark made absolutely no mention of contiguity, nor did the dissenting opinions of Judge Anzilloti or Judge *ad hoc* Vogt. The decision has essentially three bases: State activities or manifestations of sovereignty on the part of Denmark, a declaration of implied recognition of sovereignty by the Norwegian Foreign Minister and certain international agreements. It is not possible, therefore, to deduce any implied acceptance of the doctrine of contiguity by the Court.

Minquiers and Ecrehos Case (1953)

The 1953 dispute between France and the United Kingdom concerned two small groups of islands in the English Channel: the Ecrehos situated at 3.9 miles from the British island of Jersey and 6.6 miles from the French coast, and the Minquiers, at 9.8 miles from Jersey and 16.2 miles from the

French coast. Both Parties contended that they had an ancient and original title to the islands, although France often invoked the proximity and geographic dependency of the islands with the coast, and the United Kingdom occasionally referred to the natural unity of the Archipelago. At the end of its oral reply, France tried to bolster its argument on contiguity by underlining the possibility of the contested islands becoming of considerable economic importance, since it intended to build dikes between the islands and the coast so as to produce hydro-power.[88]

In its unanimous judgment awarding the islands to the United Kingdom on the basis of an ancient title, the Court makes no mention whatever of contiguity, not even of the French judge Basdevant in his separate concurring opinion. The only judge who dealt with the argument of proximity to the French coast was Levi Carneiro and he rejected it.

After remarking that the French Government had not indicated what distance from the coast constituted proximity, Judge Carneiro rejected the French argument in the following terms:

> As is stated by the French Government itself (Oral Arguments), the Minquiers and the Ecréhos are closer to Jersey than to the mainland. They must be regarded as attached to Jersey rather than to the mainland. They must be included in the archipelago. These islets were, and continue to be, part of its 'natural unity'. It is for this reason that they remained English, as did the archipelago itself.[89]

While recognizing that proximity was a relevant factor, Judge Levi Carneiro gave preference to the closer proximity of the Archipelago and to its natural unity.

Western Sahara Case (1975)

In its advisory opinion to the General Assembly concerning the legal ties between the territory of Western Sahara and the States of Morocco and Mauritania, the International Court addressed briefly the argument based on contiguity presented by Morocco. The Court rejected the argument because the alleged geographical unity was doubtful. It stated that 'the information before the Court shows that the geographical unity of Western Sahara with Morocco is somewhat debatable, which also militates against giving effect to the concept of contiguity'.[90]

The opinion of the Court adds that, even if contiguity had been present, Morocco would have had to prove a display of authority to substantiate its claim to sovereignty based on immemorial possession. The relevant passage of the opinion reads:

Even if the geographical contiguity of Western Sahara with
Morocco could be taken into account in the present connection,
it would only make the paucity of evidence of unambiguous
display of authority with respect to Western Sahara more difficult
to reconcile with Morocco's claim to immemorial possession.[91]

Contiguity of the continental shelf
North Sea continental shelf Cases (1969)
In this continental shelf delimitation dispute between three States facing
the North Sea (The Netherlands, Germany and Denmark), the Court had
to deal with the factor of adjacency which had been incorporated in the
delimitation section of the Continental Shelf Convention of 1958 (Article
1) and in its delimitation provision (Article 6). The Court's interpretation
was all the more necessary since the Truman Proclamation of 1945, which
incorporated the first claim to continental shelf jurisdiction, referred to
'submerged land which is contiguous to the continent'. The International
Law Commission, in its 1956 Report, had stated similarly that it was not
possible 'to disregard the geographical phenomenon whatever the term –
propinquity, geographical continuity, appurtenance or identity – used to
define the relationship between the submarine areas in question and the
adjacent non-submerged land'.[92]

In its judgment, the Court arrived at the conclusion that the Convention
had not used adjacency in the normal sense of the term, which would have
implied proximity. It held that there was 'no necessary, and certainly no
complete, identity between the notions of adjacency and proximity'.[93]
Indeed, it added that adjacency as used in the Convention did not prevent
a State 'from exercising continental shelf rights in respect of areas closer
to the coast of another State'.[94]

Anglo–French Channel continental shelf Case (1977)
In this continental shelf delimitation dispute between France and the
United Kingdom, the Court addressed itself to the role of 'proximity' in
determining the appurtenance of an area of continental shelf to one State
rather than to another. Having quoted from the judgment of the International Court in the 1969 cases on the meaning of proximity and adjacency,
the Arbitration Tribunal stated: 'Clearly, the Court did decline to regard
proximity as by itself a ground of title to areas of continental shelf.'[95]
However, the Tribunal continued, 'under certain conditions proximity
may be the appropriate test or method for delimiting the boundary of the
continental shelf'.[96] Having gone further than it needed to, the Tribunal
concluded that it saw 'no reason to adopt a different view of the role of

"proximity" in the circumstances of the present case'.[97] Thus we are left with a clear affirmation that proximity or contiguity, by itself, cannot be a source of title to areas of the continental shelf.

Tunisia/Libya continental shelf Case (1982)

In its examination of the location of the coast as a relevant circumstance for an equitable delimitation of the continental shelf between the Parties, the International Court stated that 'the geographic correlation between coast and submerged areas off the coast is the basis of the coastal State's legal title'.[98] At first blush, one might think that the Court had the concept of proximity in mind, but it becomes clear later that such was not the case. After quoting from the 1969 decision to the effect that it is sovereignty over the land that generates, *ipso jure,* rights over the continental shelf, the Court continued: 'As has been explained in connection with the concept of natural prolongation, the coast of the territory of the State is the decisive factor for title to submarine areas adjacent to it. Adjacency of the sea-bed to the territory of the coastal State has been the paramount criterion for determining the legal status of the submerged areas, as distinct from their delimitation . . .'.[99]

The quoted passage makes it evident that the Court was referring to the concept of adjacency interpreted in the wide sense given to it in 1969. The whole purpose of the Court's discussion of this point was to better determine how much, in terms of length, of the coasts of both Parties it should take into account in establishing the delimitation line of the continental shelf between them. In other words, while it recognized the importance of adjacency, even if taken in the normal sense of proximity or contiguity, the Court did not consider adjacency as an autonomous source of title to the continental shelf.

Gulf of Maine Case (1984)

In this case between Canada and the United States, Canada invoked the distance or proximity concept as a basis of title for the partial extension of its continental shelf and fisheries jurisdiction. The five-member Chamber of the International Court rejected the argument by stating that 'it would not be correct to say that international law recognizes the title conferred on the State by the adjacency of that shelf or that zone, *as if the mere natural fact of adjacency produced legal consequences'.* In rejecting this basis the Chamber characterized it as 'an alleged legal principle which is sometimes given the name "adjacency", sometimes "proximity" and sometimes, more especially, "distance" . . .'.[100]

3.3.   Appraisal of contiguity as a legal basis for the sector theory

This appraisal assesses the result of State practice and international decisions, summarizes doctrinal opinion and concludes with an analysis of the possibility of using the concept of contiguity as a legal basis for the sector theory in the Arctic.

State practice

State practice has revealed that the doctrine of contiguity has received a somewhat mixed acceptance as a valid basis for a claim to territorial sovereignty. Among Arctic States, the doctrine has received some support from Canada and the Soviet Union, but has been opposed by the United States, Norway and Denmark.

In the Antarctic, it seems to have been invoked by Great Britain and Chile but only along with effective occupation. Most claimant States, however, might be said to rely on the closely related hinterland theory, in the sense that their occupation of the coastal territory is the main basis for their claim to the whole sector ending at the South Pole. This was admitted by Great Britain in 1929. Nevertheless, Norway is an exception and, because of its strong opposition to contiguity and the sector theory, it has refused to project its claim as far as the South Pole.

International decisions

In spite of the above State practice, part of which is favourable to contiguity, international decisions have accorded it a rather limited and qualified acceptance. It was expressly rejected by the arbitrators in the island cases of *Aves* (1965), *Monks* (1885) and *Palmas* (1928), and was implicitly rejected by the International Court in the *Eastern Greenland* (1933) and *Minquiers and Ecrehos* (1953) *Cases*.

In fact, it appears that the only actual application of the doctrine can be found in the *Bulama Island Case* (1870), where the arbitrator awarded the island to Portugal because the latter had discovered it and it was situated within the territorial waters of the coastal possession of that country. The application of contiguity to that situation was approved in a dictum by the arbitrator in the *Island of Palmas Case* (1928), but he was careful to specify that this application could not be extended to islands outside territorial waters. In that case also, arbitrator Max Huber admitted that contiguity could serve as a basis for the acquisition of sovereignty over a group of islands which could be considered as a unit. However, he added the important limitation that such contiguity could not dispense with the necessity of the usual display of State activities for the maintenance of the already acquired sovereignty.

The possibility of contiguity conferring a preferential right for the acquisition of sovereignty over the whole of a region constituting a single organic whole was expressed in a dictum by the arbitrator in the *British Guyana Case* (1904). However, such preferential right cannot extend to the actual presumption of sovereignty, as stated by Max Huber in his landmark decision of 1933, since this would be contrary to the duty to display activities of a State. A somewhat similar view was expressed by the International Court in the *Western Sahara Case* (1975) when it stated that contiguity was of little assistance when immemorial possession was relied upon as a basis for a claim to sovereignty.

The only other international decisions where contiguity has been given consideration are in three cases of continental shelf delimitation. In none of those cases was contiguity held to constitute a source of title to the resources of submarine areas. In the 1969 *North Sea Case*, the term adjacency was held to be different than proximity; in the *Anglo–French Channel Case* of 1977, the Tribunal declined to give any role to proximity as a source of title; in the *Tunisia/Libya Case* of 1982, the Court used proximity only as a relevant circumstance to determine the length of coasts which should be considered in the establishment of the delimitation line and in the *Gulf of Maine Case* of 1984, the Chamber of the Court rejected completely proximity as a basis for continental shelf or fisheries jurisdiction. Distance from the coast was recognized by the International Court in the *Malta/Libya Case* of 1985 as a legal basis for continental shelf rights, but only within the new concept of the 200-mile exclusive economic zone.

In only one of the three continental shelf decisions was there a concurring judge who addressed himself to contiguity as a basis to justify 'territorial or maritime acquisitions'. After quoting with approval from the decision of Max Huber in the *Island of Palmas Case*, Judge Ammoun referred specifically to the Arctic and Antarctic sector claims when he rejected the contiguity argument in the following passage:

> . . . it would not be superfluous to stress the seriousness of the consequences which the acceptance of this argument would involve. It would justify territorial or maritime acquisitions repugnant to the fundamental principles of contemporary international law: for example the appropriation of large areas of the Arctic Ocean and the Antarctic Continent, an appropriation which also relies on the doctrine of sectors, which doctrine, in certain of its elements, is reminiscent of the abandoned concept of *spheres of influence* . . . .[101]

This rejection of the doctrine of contiguity with special reference to the sector theory in the polar regions constitutes the only such pronouncement from a judge of the International Court.

Doctrinal opinion

By the beginning of the twentieth century, the doctrinal opinion on the role of contiguity had evolved along the same lines as State practice and international jurisprudence, namely toward a rejection of the concept as a basis for the acquisition of title. Writing in 1925, the great French jurist Fauchille reviewed such bases as contiguity used by States to circumvent the difficulties in meeting the conditions of effective occupation. He concluded as follows:

> Toutes ces solutions, qui dérivent de la mise en oeuvre du droit de continuité, doivent être absolument repoussées, car elles aboutissent en dernière analyse à ne tenir aucun compte du principe qu'une occupation, pour créer un titre de possession exclusive, doit être effective, c'est-à-dire non-fictive.[102]

Before 1925, the role of the contiguity doctrine had been generally studied within the context of temperate zones. With the advent of sector claims by Canada and the Soviet Union in the Arctic and those of Great Britain (on behalf of itself in 1908 and of New Zealand in 1923) in the Antarctic, writers began to examine contiguity in the context of polar regions.

In his 1925 article the American jurist, D. H. Miller, referred specifically to Canada's claim, and expressed his qualified support for it in the following terms:

> It cannot be said, however, that such a claim as this is wholly without foundation or precedent. It bears some analogy to the 'back country' or 'hinterland' theory regarding territory stretching away from the coast. More accurately, it may be said to rest partly on the notion of 'territorial propinquity' which the United States on one famous occasion recognized as creating 'special relations between countries'.[103]

The 'famous occasion' presumably refers to the 1917 agreement between the United States and Japan which stated that 'territorial propinquity creates special relations between countries' and led to the United States' recognition of Japanese special interests in China, particularly in the region contiguous to its possessions.[104] Miller, however, would not go so far as to say that contiguity existed as a legal principle. He confined himself to stating that 'claims to unoccupied territory on the ground of contiguity are not unknown, although it cannot be said that there is any well-defined or clearly settled principle to support them'.[105]

In his 1926 major study on 'The Acquisition and Government of Backward Territory in International law', the English writer, M. F. Lindley, could not express a favourable opinion as to the legal role of contiguity in general or in the polar regions in particular. After a searching analysis of the doctrine, and having referred to it specifically as its basis for Canada's sector claim, he concluded that 'it does not appear, however, that the doctrine has acquired any greater legal sanction in its application to those regions than in regard to unoccupied territory in other parts of the world . . .'.[106] In short, Lindley's position was that 'simple geographic contiguity is not a source of legal rights,' while admitting that it may have 'important political consequences'.[107]

In 1930, the Soviet writer and main protagonist of the sector theory, W. L. Lakhtine, invoked a number of contiguity-related concepts as substitutes for effective occupation. 'It has become evident', he wrote, 'that in polar regions "effective occupation" cannot be realized and a substitute principle that sovereignty ought to attach to littoral States according to "region of attraction" is now suggested and practically applied.'[108] He then applied his theory of regions of attraction and divided the Arctic into sectors.

Contiguity has since been discussed by many writers in its possible application to the polar regions and as a legal basis for sector claims. Although some of them are prepared to recognize a certain role for contiguity in the course of applying the principle of effective occupation, virtually all of them reject it as a proper independent source of title. Before Waldock's excellent study in 1948, to which reference will be made shortly, the following writers had concluded that there should be either a complete or a qualified rejection of the doctrine: Smedal (1932),[109] Strupp (1933),[110] Hyde (1934),[111] von der Heydte (1935),[112] and Gidel (1948).[113]

In his article of 1948 on the Falkland Islands Dependencies, Professor Waldock made a thorough analysis of the hinterland and contiguity doctrines as possible roots of title, in light of the *Island of Palmas* and *East Greenland Cases*, and expressed his conclusion as follows:

> It is not believed that either form of sector doctrine can by itself be a sufficient legal root of title. The hinterland and contiguity doctrines as well as other geographical doctrines were much in vogue in the nineteenth century. They were invoked primarily to mark out areas claimed for future occupation. But, by the end of the century, international law had decisively rejected geographi-

cal doctrines as distinct legal roots of title and had made effective occupation the sole test of the establishment of title to new lands. Geographical proximity, together with other geographical considerations, is certainly relevant, but as a fact assisting the determination of the limits of an effective occupation, not as an independent source of title.[114]

Professor Waldock then quoted with complete approval the passages already reproduced from the Award of Max Huber in the *Island of Palmas Case* and added that 'the judgment of the Permanent Court in the *Eastern Greenland Case* does not, it is submitted, conflict at all with the views of Judge Huber as to the non-legal character of the proximity doctrines'.[115]

Professor Waldock concluded his opinion on Antarctic sector claims based on contiguity by saying:

> If the above appreciation of the place of continuity and contiguity in international law is correct, sector claims in the Antarctic, being merely forms of continuity or contiguity, can have no legal significance independently of an exercise or display of state activity in regard to the sector.[116]

Since 1948, doctrinal opinion has remained basically unchanged in its opposition to contiguity as an independent source of title to polar region sectors in general and those in the Arctic in particular. The following constitute an impressive and representative group: G. W. Smith (1952),[117] C. M. Franklin and V. C. McClintock (1952),[118] O. Svarlien (1960),[119] F. M. Auburn (1972),[120] and G. D. Triggs (1983).[121] An important study was made in 1950 by Professor H. Lauterpacht in which he addressed the question of contiguity as the legal basis of sovereignty over submarine areas; the learned jurist gave considerable weight to the element of contiguity, but he concluded that the legal source of title had to be found also in 'other factors such as general acquiescence'.[122]

3.4.   Summary

Our appraisal leads to the definite conclusion that, as a general rule, contiguity and its related geographical doctrines are not sufficient in themselves to serve as a legal basis for the acquisition of territorial sovereignty. There are two possible exceptions: an island in the territorial sea of a coastal State and an island sufficiently linked with a group of islands or archipelago as to form part of it. This would apply particularly to remote areas such as the Arctic regions.

In addition, contiguity is a relevant consideration when applying the principle of effective occupation for the determination of sovereignty over vast areas of continents or large islands. It is not altogether clear whether such contiguity results simply in a preferential right for the acquisition of sovereignty or in an actual presumption of sovereignty. It is more logical and consonant with international order that it be the latter.

Contiguity has also been held to play a role in attaining an equitable delimitation of the continental shelf between neighbouring States. Although contiguity by itself is not a source of title, it is a relevant circumstance to determine the length of coasts which should be considered in the establishment of the line of delimitation for the continental shelf appertaining to the land territory of each State.

The question remains as to what legal effect contiguity may have with respect to the acquisition of sovereignty over sea areas. Contiguity by itself, being unable to generate sovereignty over land areas, *a fortiori* cannot lead to the acquisition of sovereignty over sea areas. Only territorial sovereignty can, *ipso jure,* generate rights over adjacent waters. Nevertheless, the concept of contiguity (in an extended meaning) has now become the legal basis for the exclusive jurisdiction of a State over the natural resources within a 200-mile zone from the coast. The resources include those of the continental shelf, if any, and of the water column.[123] Of course, sovereignty over the land territory remains a condition precedent for the generation of such jurisdiction and the latter is not tantamount to sovereignty, since the traditional freedoms of the high seas remain. If sovereignty in the zone were the result, contiguity (in this extended meaning) would really generate sovereignty.

The inescapable conclusion, since contiguity does not constitute an autonomous source of sovereignty over land areas (subject to the qualifications indicated) or over maritime areas beyond the territorial sea, must be that *contiguity is incapable of serving as a legal basis for the sector theory.*

4

Customary law as a basis for the sector theory

In spite of the conclusion that the sector theory cannot find a valid legal basis in the concept of contiguity, the practice of States might be such that the theory has resulted in a principle of customary law. This would appear to be the view held by a small number of authoritative commentators. For instance, Professor J. S. Reeves wrote, in a brief comment on Antarctic sectors in 1939, that '[o]ne may assert that the sector principle as applied at least to Antarctica is now part of the accepted international legal order'.[124] Numerous references to State practice are also found in the writings of proponents of the sector theory, such as David Hunter Miller and W. L. Lakhtine.

The above statement by Professor Reeves did not go unchallenged and, in 1948, Professor Waldock expressed his disagreement with the notion that State practice had resulted in a new rule of international law. Having referred to Professor Reeves' statement, he wrote:

> It is, however, scarcely possible to regard state practice as sufficiently certain and general to establish a new rule of international law. The United States, a potential claimant in both the Arctic and Antarctic, has consistently denied that sector claims have any legal force. Neither Norway nor Denmark has claimed Arctic sectors. Nor have non-sector states given any recognition to the legality of sectors. On the contrary, Nazi Germany was showing signs of claiming part of Norway's sector and Japan entered a reservation on hearing of Chile's declaration. Moreover, Chile and Argentina are contesting the United Kingdom's sector.[125]

Some 35 years having passed since Professor Waldock's examination of State practice, it is now advisable to re-examine the question. After a

brief recall as to the requirements of customary law, this chapter reviews
State practice in the Arctic and Antarctic.

4.1.   Requirements of customary law

As specified in the Statute of the Court, what is necessary to
establish a principle of international customary law is 'a general practice
accepted as law.'[126] Two elements must be shown to exist: first, a material
element, that is a practice followed by the generality of States; and,
second, a psychological element, namely the acceptance of such a practice
as legally binding.

As to the first element, it is sufficient to show the instances where
States, particularly those directly affected by the alleged custom, have in
fact followed the practice in question. As for the second element,
however, it must be shown that States followed such a practice because
they felt under a legal obligation to do so. These two requirements were
expressed by the International Court as follows:

> Not only must the acts concerned amount to a settled practice,
> but they must also be such, or be carried out in such a way, as to
> be evidence of a belief that this practice is rendered obligatory by
> the existence of a rule of law requiring it. The need for such a
> belief, i.e., the existence of a subjective element, is implicit in the
> very notion of the *opinio juris sive necessitatis*. The States con-
> cerned must therefore feel that they are conforming to what
> amounts to a legal obligation. The frequency, or even habitual
> character of the acts, is not in itself enough.[127]

The above requirements are those of an ordinary general custom.

If a party relies on a *regional custom,* that is one which applies only in a
certain geographical region, the requirements are the same except that the
burden of proof is slightly more stringent. Colombia was faced with this
heavier burden of proof in the *Asylum Case* in 1950, when it invoked
'American international law', that is, a regional custom peculiar to Latin
American States in relation to the granting of asylum. The Court described
the burden of proof incumbent on Colombia in the following terms:

> The Party which relies on a custom of this kind must prove that
> this custom is established in such a manner that it has become
> binding on the other Party. The Colombian Government must
> prove that the rule invoked by it is in accordance with a constant
> and uniform usage practised by the States in question, and that
> this usage is the expression of a right appertaining to the State
> granting asylum and a duty incumbent on the territorial State.[128]

In that case, the Court came to the conclusion that the practice of granting asylum among Latin American States had been 'so much influenced by considerations of political expediency in the various cases, that it is not possible to discern in all this any constant and uniform usage, accepted as law, with regard to the alleged rule of unilateral and definitive qualification of the offence'.[129]

The question arises here whether we are dealing with a regional custom or a general one. Taking the strict view that we are only concerned with the Arctic, the alleged custom would have to be a regional one. However, presuming that the sector theory has been applied to both polar regions and taking into account the considerable number of States involved, both possibilities are examined.

4.2.   State practice in the Arctic

As much as it can be determined from public documents, the practice of the five Arctic States is reviewed. In order to be as complete as possible, even indirect and partial uses of the sector theory are included. This is particularly so with respect to Canada, for which information is more readily available.

State practice of Canada

Canada has made use of the sector theory, in whole or in part, on a number of occasions.

Taking of possession based on sector theory (1909)

In 1909, the year Robert E. Peary reached the North Pole on April 6, Captain J. E. Bernier, heading his second expedition to the Arctic (1908–9), took formal possession of the whole Arctic Archipelago on July 1 by unveiling and depositing a plaque at Parry's Rock on Melville Island. The inscription on the plaque read as follows:

> This Memorial is erected today to commemorate the taking possession for the DOMINION OF CANADA of the whole ARCTIC ARCHIPELAGO lying to the north of America from long. 60° W to 141° W up to latitude 90° N. Winter har. Melville Island. CGS Arctic. July 1st 1909. J. E. Bernier. Commander.[130]

The above inscription describes, of course, the full sector shown on the map of 1904 and claimed for Canada by Poirier in his resolution of 1907.

Apparent abandonment of the sector theory (1910-24)

On February 1, 1910, Senator Poirier asked if the government intended to appoint a commissioner to supervise the lands and islands in the Arctic upon which the Canadian flag had recently been hoisted.[131] He made a long speech in support of his question, during which he stated that Canada was entitled to claim all Arctic lands situated between Hudson Bay and the recently discovered Pole.[132] Poirier's question was virtually ignored and it would appear that, from 1910 to 1924, no direct reference to the sector theory was made in government circles. Indeed, there is evidence that in the early 1920s Canada decided to abandon the sector theory.

Captain Bernier was sent to the Arctic in 1910 to patrol the northern waters and affirm Canada's jurisdiction but, in 1913, when Prime Minister Borden learned that Stefansson's expedition to the Canadian Arctic was to be partly financed by the National Geographic Society, he wrote to its director stating that it would be more appropriate if all of the expenses of the expedition were to be paid by the Canadian Government.[133] It was during that expedition, which lasted some five years (1913–18), that Stefansson discovered Brock and Borden Islands north of Prince Patrick Island on the edge of the Beaufort Sea. It was also during the early part of this expedition that the party left by Stefansson in the *Karluk* became icebound and eventually drifted toward Wrangel Island north of Siberia. A few members of the group managed to reach the island by foot and stayed from March 12 to September 7, 1914[134] during which time they seemed to have taken possession of the island for Canada.[135]

The year 1920 marks a point of definite uncertainty on the part of Canada as to whether or not it ought to confine its territorial claim to the so-called Canadian sector. On the one hand, Canada was anxious to insure that its claim of sovereignty over Ellesmere Island would not be challenged by Denmark and, on the other, it was being pressed by Stefansson to occupy Wrangel Island and establish a Canadian fur company there in order to support the validity of its claim. As for Ellesmere Island, the Government had written to Denmark on July 31, 1919, asking the Danish Government to restrain Greenland Eskimos from killing musk oxen on the island.[136] In its reply of April 26, 1920, the Danish Government stated that it thought it could subscribe to the view expressed in a report by the Danish explorer Rasmussen that Ellesmere Island was 'no man's land.'[137] This resulted in an appropriate communication being sent to Denmark by Great Britain, on behalf of and at the request of Canada, and Denmark did not pursue the matter after that. The fact that Great Britain recognized Denmark's claim to sovereignty over Greenland at about the same time might also have helped to convince Denmark not to make any claim relating to Ellesmere Island.

In a memorandum dated October 28, 1920 entitled 'Exploration and Occupation of the Northern Arctic Islands', the legal adviser to the Prime

Minister recommended a number of steps which should be taken to confirm Canada's sovereignty over Ellesmere Island, including a Canadian Government expedition to be headed by Stefansson. All this, however, was not to be taken as a reliance on the sector theory since, in the same memorandum, the legal adviser stated that there might be advantage in encouraging the settlement of Wrangel Island by some Canadian company, such as the Hudson's Bay Company. And, in the last paragraph, there is the following submission:

> It is also submitted that in the future we should refrain in official or public documents from admitting that the 141st meridian north of Alaska constitutes the western boundary of the Canadian domain. Official documents in the past have implied such an admission. There is no need for this. The treaty defining the Alaska boundary carried the 141st meridian only to the frozen ocean.[138]

The suggestion of claiming Wrangel Island did not find favour with the members of the Department of the Interior and with the Under-Secretary of State for External Affairs, Joseph Pope, who personally preferred effective occupation.[139]

In spite of the above, Prime Minister Meighen wrote to Stefansson on February 19, 1921, declaring that the Canadian government proposed to press its claim to Wrangel Island.[140] However, official government approval was never given, and Stefansson decided to organize a private expedition with the intention of obtaining the necessary government approval eventually. So, on September 16, 1921, Stefansson's expedition, headed by one Alan Crawford of Toronto, raised the Canadian flag as well as the British flag and took possession of Wrangel Island in the name of the King of England.[141]

The claim to Wrangel Island was affirmed the next year by the Mackenzie King government. On May 12, the Prime Minister himself, in answer to questions from members of the Opposition, including Meighen, stated that the government certainly affirmed that Wrangel Island was part of Canadian territory.[142]

Return to the sector theory (1924–5)
In 1924, the year after the official agent of the Soviet Government in Great Britian had objected to the taking of possession of Wrangel Island by a group of Canadians, the Minister of the Interior, Charles Stewart, was asked to explain the situation with respect to Wrangel Island. He replied that the government had told Stefansson that it was not interested

in Wrangel Island, that it did not know officially what position the British government would take as to a possible claim for itself, but concluded that 'so far as Canada is concerned, we do not intend to set up any claim to the island'.[143]

In the same year, the Department of the Interior published a map of the Northwest Territories which showed the traditional sector lines right up to the Pole, which seemed to indicate an intention on the part of government to revert to the sector theory (see Figure 5).

On June 1, 1925, the Minister of the Interior, Charles Stewart, moved the second reading of an amendment to the *Northwest Territories Act,* providing for the issuing of licences and permits to scientists and explorers who wished to enter the Northwest Territories. In answer to a question as

Fig. 5. Northwest Territories, 1924. *Source*: Public Archives of Canada.

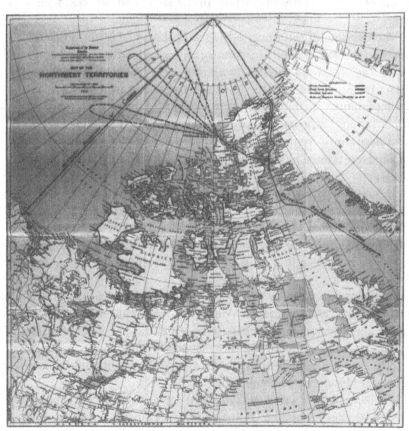

to whether Canada claimed sovereignty right up to the North Pole, the Minister answered: 'Yes, right up to the North Pole'.[144]

This, of course, represented a return to the sector theory and was not something which the Minister had decided to do on the spur of the moment. Indeed, the government had been considering for some time what action to take in light of the United States project, in 1924, to send the navy's dirigible, the *Shenandoah,* to the Arctic. Although the project had been abandoned, it had been replaced by another one, led jointly by Commander Richard E. Byrd and Dr Donald B. MacMillan. The main purpose of the expedition was to verify the possible existence of 'Crocker Land', which Commander Peary thought he had seen in the Arctic Ocean, north of Canada. As stated by the Honourable Charles Stewart, '[h]ere we are getting after men like MacMillan and Doctor Amundsen, men who aré going in presumably for exploration purposes, but possibly *there may arise a question as to the sovereignty over some land they may discover* in the northern portion of Canada, and we claim all that portion.'[145]

The question of returning to the sector theory had been studied by a special Interdepartmental Committee two weeks earlier, and the advantages and disadvantages had been debated. The spokesman for the Department of the Interior favoured the sector theory, whereas the representatives of the Department of Justice and of External Affairs preferred to rely on occupation and control of the islands in question.[146] The decision to resume claiming on the basis of the sector theory was made even clearer later on, during the exchange in the House, when Stewart stated: '[w]e have no interest in Wrangel Island, and the British government have expressed themselves to the same effect.'[147]

Notes were exchanged between Canada and the United States during the months of June and July 1925, in the course of which Canada insisted on the necessity for the MacMillan expedition to obtain a permit pursuant to the recent legislation. In point of fact, the MacMillan expedition left Boston on June 17 and did not obtain any permit, in spite of a statement to the contrary made by Commander Byrd to Canada's former Gold Commissioner Mackenzie on August 19, at Etah, North Greenland, aboard the Canadian patrol ship *Arctic* under the command of Commissioner Mackenzie. Following this meeting, during which Commander Byrd refused Mackenzie's offer to issue him with a licence, the flights over Ellesmere Island were abandoned.[148]

On December 9, 1925, the Governor General sent a dispatch to the British Ambassador to the United States reviewing the facts surrounding this incident and ending with the following request:

I would request Your Excellency to have the goodness to draw the attention of the United States Secretary of State to the apparent failure on the part of the expedition to observe the requirements of the Canadian laws. I enclose, for convenience of reference, copies of the laws in question, together with copies of three affidavits taken by Messrs MacKenzie, Morin and Steele.[149]

Arctic Islands Preserve based on sector theory (1926–9)
On July 19, 1926, a Canadian Order in Council was adopted, creating the *Arctic Islands Preserve* and describing the northern part of the Preserve in sector form, going as far as the North Pole. The description reads: 'comprising all that tract of land which may be described as follows:' [then follows a description ending thus]: '. . . on the west to its intersection with the meridian of longitude 60° West; thence, due north, following said meridian to the North Pole; thence, due south, following the 141st meridian to latitude 75° North . . .'.[150]

A memorandum dated June 23, 1926, signed by Mr Finnie and addressed to Deputy Minister Cory in support of the Order establishing the Preserve, states that '[t]he whole purpose of the memorandum is to protect both the natives and the wildlife and to place something on the map to indicate that this government control and administer the area between the 65th *(sic)*, and 141st degrees of longitude right up to the Pole.'[151] The memorandum also refers to a map which is attached and 'which outlines in blue, the area covered by the proposed Arctic Islands Preserve'.[152]

On May 15, 1929, another Order in Council was adopted establishing new Game Regulations and describing the Arctic Islands Preserve in the same sector form as did the one in 1926.[153]

In 1929, the Department of the Interior published a 'Map of the Northwest Territories' showing four preserves, one of which being the Arctic Islands Preserve, with boundaries and names of all preserves in red (see Figure 6).

Further reliance on sector theory (1938–54)
On May 20, 1938, the Minister of Mines and Resources, the Honourable Mr Crerar, was asked if there was not an international recognition that Canada had 'a sort of sovereignty over the triangle bounded by the Alaska–Yukon line, continued to the Pole, and then to the west coast of Greenland'. He replied: 'Yes. What is known as the sector principle, in the determination of these areas is now very generally recognized, and *on the basis of that principle as well our sovereignty extends right to the Pole*

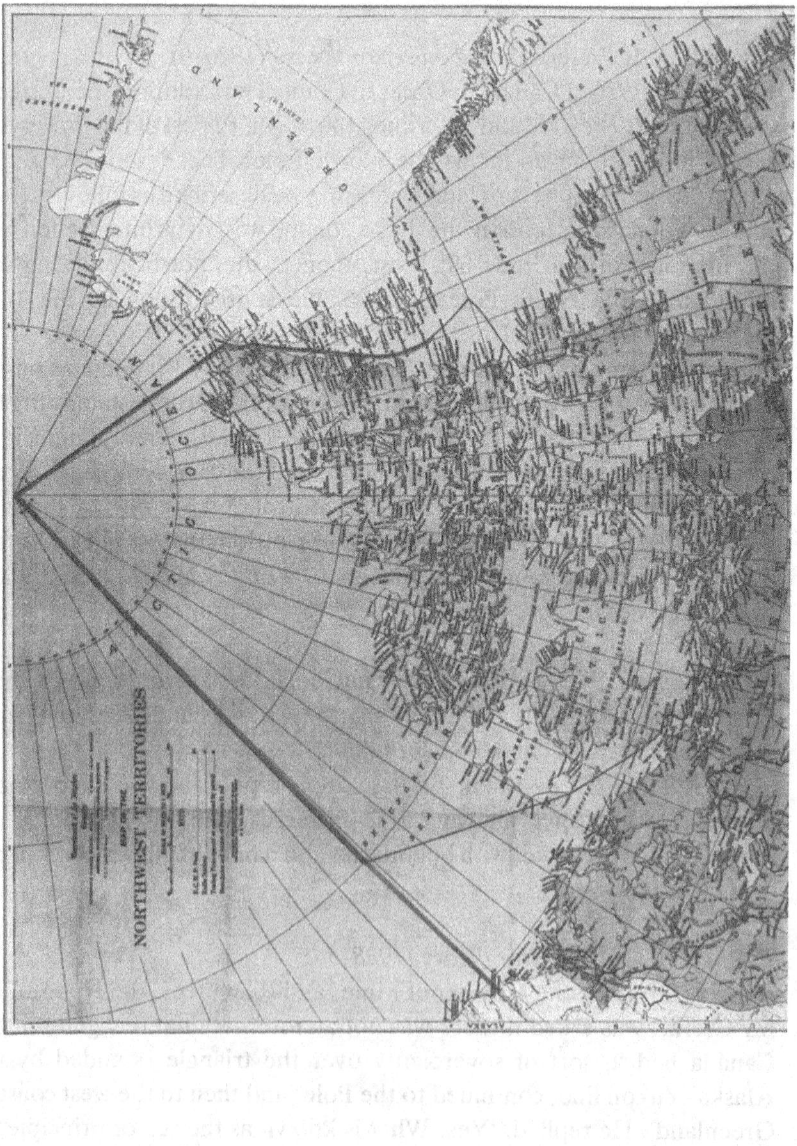

Fig. 6. Northwest Territories (Arctic Islands Preserve), 1929. *Source:* Public Archives of Canada.

within the limits of the sector'.[154] On the same occasion, when the Minister was asked if Wrangel Island was Canadian territory, he replied: 'No, Wrangel Island is Russian territory'.[155]

In 1939, the Department of Mines and Resources published a map entitled 'Northwest Territories and Yukon' showing the Arctic Islands Preserve, as had been done on the 1929 map, right up to the North Pole.

On September 20, 1945, the Arctic Islands Preserve was extended in a southward direction and the northern limits were still described in sector form right up to the North Pole.[156]

During the year 1946, the United States made two requests to the Canadian government. One was for the U.S. Navy to carry out certain training exercises in the waters of the Canadian Arctic, and the other was to install an Arctic weather station. In considering those two requests (the first of which was granted and the second left in abeyance), government officials, concerned about Canada's sovereignty, discussed the validity of the sector theory. In a memorandum from the Department of External Affairs to the Cabinet Defence Committee on May 30, 1946, the fear was expressed that, if the United States were allowed to carry out their Arctic weather station programme without Canadian participation, Canadian sovereignty might be endangered by: '(a) claims by the United States to *territory which is assumed to belong to Canada under the Sector Theory;* (b) claims to air rights on the basis of development: such claims were widely voiced in the United States when that country built air fields in Canada during the war'.[157]

Upon receipt of the above memorandum, the Canadian Ambassador to the United States, Lester B. Pearson, thought that Canada should take advantage of the situation and secure U.S. recognition of the sector principle. On June 5, 1946, he wrote to the Acting Under-Secretary of State for External Affairs:

> I am wondering whether we could not take advantage of the present situation to secure from the United States Government public recognition of our sovereignty of the total area above our northern coasts, based on the *sector principle.* The memorandum feels that this is not necessary, and even possibly inadvisable, because insistence on a formal assurance of respect for Canadian sovereignty might indicate doubt on our side of the validity of our claim to such sovereignty. Without attempting to insist on any-thing, I think we might persuade the United States authorities that it would be in their own interests at this time to reinforce our claim to the area under the sector principle.[158]

A subsequent letter by Pearson, dated June 18, indicates that the suggestion was not very well received in Ottawa. His letter states: 'I fully appreciate the difficulties and possible disadvantages in the way of carrying out this suggestion, and in view of what you say I suppose it would be inadvisable to proceed even along the lines suggested in my earlier letter'.[159]

In July 1946, Pearson published an article in *Foreign Affairs* entitled 'Canada Looks Down North' in which he advanced the sector theory. He described Canada in the following terms:

> A large part of the world's total Arctic area is Canadian. One should know exactly what this part comprises. It includes not only Canada's northern mainland but the islands and the *frozen sea north of the mainland* between the meridians of its east and west boundaries, extended to the North Pole.[160]

Strickly speaking, this position cannot be equated to that of the Canadian government since Ambassador Pearson was, of course, writing in his personal name. However, he was identified in the publication as Canada's ambassador to the United States, and he was expressing himself in what might be called a representative type of language, particularly when he wrote: 'The Canadian Government, while ready to cooperate to the fullest extent with the United States and other countries in the development of the whole Arctic, accepts responsibility for its own sector'.[161]

In 1951, the Department of Mines and Technical Surveys published a new map of Canada which showed Canada's boundary as following the sector right up to the Pole (see Figure 7). In 1951, the map was described by Pearson, who had become Under-Secretary of State for External Affairs, as representing Canada's 'new look'. He wrote:

> These northern defence activities and scientific projects have given Canada a 'new look.' This new look was for the first time reflected last year when the Department of Mines and Technical Surveys of the Canadian government published a new map of Canada for use by the general public. Although it is an ordinary political and administrative map of the scale of 100 miles to the inch, its publication was something of an event, because it was the first time, in a publication for general distribution, that all of Canada has been shown.[162]

On December 8, 1953, in moving the second reading of a bill establishing the Department of Northern Affairs and national resources, Prime Minister St Laurent stressed the growing importance of the north and stated: 'We must leave no doubt about our active occupation and exercise of our *sovereignty in these northern lands right up to the Pole'*.[163]

In 1954, the *Canada Yearbook,* published by the Department of
Commerce, described Canada as extending to the North Pole and comprising the Arctic Archipelago between the 60th and the 141st meridians.[164]

Sector theory restricted to land (1955–7)

In March 1955, the Deputy Minister of the Department of Northern
Affairs and National Resources, R. G. Robertson, made a clear statement on the sector theory before the Special Committee on Estimates. At
the request of his Minister, Jean Lesage, he answered Mr Thatcher's
question as to the presence of a Soviet ice island:

> I assume Mr Thatcher is referring to the indication that these
> Soviet scientists floated into what is sometimes called the Cana-
> dian sector in the north. *Canada has never formally asserted a*

Fig. 7. Map of Canada (with sector lines), 1951. *Source:* Public Archives of
Canada.

claim to the northern sector as such. Sector lines have been drawn on the map since about 1903 at which time there was no complete knowledge of the land that is in the far north and the indication was that *Canada was, in effect, claiming any land within this sector line,* though there was no formal statement of claim.[165]

Mr Robertson added:

The sector lines were not drawn up, however, to indicate any claim to water or ice, so when this ice island floated into that sector where it is all water, it entered an area to which there has never been a Canadian claim formally extended.[166]

On August 3, 1956, a statement on the sector theory was made by Jean Lesage, the Minister for the new Department of Northern Affairs. In answering a question on whether Canada claimed ownership to the ice 'cap' north of the land area on the basis of the sector principle, Mr Lesage replied:

We have never subscribed to the sector theory in application to the ice. We are content that our sovereignty exists over all the Arctic islands . . . *We have never upheld a general sector theory.* To our mind, the sea, be it frozen or in its natural liquid state, is the sea; and *our sovereignty exists over the lands and over our territorial waters.*[167]

Being pressed with further questions by the Opposition, Mr Lesage repeated three times that the ice-covered area north of the islands was high seas.[168]

On March 7, 1957, when Mr Lesage was accused by a member of the Opposition (Mr Harkness) of having disavowed the sector theory, which he said had been maintained by a former Minister of the Interior, he replied: 'What I said was that Canada has claimed, for a great number of years, *all land* within that sector as being Canadian territory.'[169] He added later on: 'I said that we did not claim the open seas in either liquid or frozen form, but that we claim as Canadian *all land* within that sector up to the North Pole'.[170]

Sector theory seemingly applied to water areas (1957–8)
After a change in government the same year, Lesage, then in the Opposition, put a written question to the government, asking if the waters of the Arctic Ocean from the Arctic islands to the North Pole within the so-called 'Canadian sector' were Canadian waters. A week later, the new Minister for Northern Affairs and Natural Resources, Alvin Hamilton, gave a rather long, non-committal answer during which he stated in

particular: 'It will be seen, then, that the Arctic Ocean north of the Archipelago is not open water nor has it the stable qualities of land. Consequently, the ordinary rules of international law may or may not have application'.[171]

In 1958, the sector theory received an appreciable degree of attention in the House of Commons, due mainly to the new Leader of the Opposition, Lester B. Pearson. On July 7, 1958, he made the following statement:

> Indeed, we have claimed sovereignty under what we call the *sector theory* over the prolongation north, right to their meeting at the North Pole, of the east and west extensions of our boundary. If we are to make that claim stick, and it certainly is important we should make it stick, we have to do everything that is possible, everything that is practical, to develop those areas and reinforce whatever rights we may have in law with the right of occupation.[172]

On August 14, 1958, Pearson referred to the passage of two American submarines under polar ice which, according to the application of 'a particular theory of international law', is considered Canadian territory.[173] In the same speech, Pearson referred to previous claims made by Senator Poirier in 1907, Charles Stewart in 1925, and Prime Minister St Laurent in 1953. He went on to suggest to the Minister of Northern Affairs, however, that perhaps the sector theory would not suffice as a basis for Canada's sovereignty claim. He said:

> I suggest to the minister that we shall have considerable difficulty in doing that if we base our claim merely on the sector theory which has not yet, I think, been generally considered a valid doctrine in international law.
> *The sector theory itself is not enough; it must be followed by rights based on discovery and effective occupation.* That had been very much in the mind of the previous government, as no doubt it is in the mind of the present government, to buttress any claims we have under one theory of international law by rights of discovery and effective occupation, which are acknowledged by all as valid in international law.[174]

In his reply, Hamilton, the Minister of Northern Affairs, stated: 'I am very pleased that the Leader of the Opposition has accepted the phrase, "effective occupation", because you can hold a territory by right of discovery or by claiming it under some sector theory, but where you have great powers holding different points of view, the only way to hold that territory, with all its great potential wealth, is by effective occupation'.[175]

A couple of days later, on August 16, 1958, in answer to a question by Mr Pearson as to the procedures which had to be observed by Canadians visiting United States defence installations in the Canadian Arctic, Prime Minister Diefenbaker gave the following reply:

> When I became Prime Minister, one of my first acts was to have this question looked into in detail, with a view to assuring that while we cooperate in defence willingly and freely, in no way shall our sovereignty be impeded or interfered with; and further than that, that everything that could possibly be done should be done to ensure that *our sovereignty to the North Pole be asserted,* and continually asserted, by Canada.[176]

Sector theory applied to islands (1964)

On September 4, 1964, in answer to a question as to whether the Government of the United States recognized Canada's title to all islands north of the mainland up to the North Pole, the Minister of Northern Affairs and Natural Resources, Arthur Laing, replied as follows:

> For some years now, the present government and the previous governments have asserted our *sovereignty over the islands* extending northward from the meridians appropriate to Canada: on the west from the meridian which divided Alaska from the Yukon and on the east from the appropriate eastern meridian, with a sinuosity devised to provide for the existence of Greenland and thence extending to the North Pole. Canada has not only asserted its sovereignty there but we are implementing our sovereignty in many places. I have never heard or seen that assertion called in question by any nation, let alone the United States of America.[177]

Although there is a reference to the North Pole, it is obvious that it relates to the meridians only and that the sector theory is used merely to delimit the islands over which Canada claims sovereignty.

Sector theory applied to continental shelf (1965–9)

In January 1965, the Department of Northern Affairs and Natural Resources began to issue oil and gas exploration permits in the Beaufort Sea up to and along the 141st meridian. In the same year, the Department of Energy, Mines and Resources sent to the United States Department of the Interior a small-scale map indicating the offshore areas where oil and gas permits had been issued.

On March 10, 1969, the sector theory was the subject of an exchange between the Rt Hon. John Diefenbaker and Prime Minister Trudeau. Diefenbaker asked the following question:

> Does the Prime Minister accept the sector principle and, if he does, and if that is still the policy of the Government of Canada, then waterways are in the same position as islands and lands?[178]

Prime Minister Trudeau replied:

> Mr Speaker, I do not have the same understanding of this as the Rt Hon. gentleman. *I believe the sector theory applies to the seabed and the shelf. It does not apply to the waters.* The continental shelf is, of course, under Canadian sovereignty – this is the seabed, but not the waters over the shelf.[179]

Later, in the exchange, the Prime Minister repeated his assertion that the sector theory would apply to the seabed but not to the waters or ice. He added that 'the question of whether the waterways and the frozen ice are in the international high seas, in territorial waters or in inland waters is the question that is in dispute'.[180]

Implied rejection of sector theory by Canadian Court (1969)

In 1969, an implied rejection of the sector theory was made in the Territorial Court of the Northwest Territories by Mr Justice Morrow in the case of *Regina v. Tootalik, E4-321.* An Eskimo had been accused of unlawfully hunting on offshore ice some 10½ miles from Boothia Peninsula, contrary to a game ordinance adopted by the Commissioner in Council of the Northwest Territories. The judge came to the conclusion that the term 'territories', defined in the Northwest Territories Act as being 'all that part of Canada north of the 60th parallel of north latitude', included the sea ice in question.[181] The basis for his opinion was that the transfer of title from Great Britian to Canada, by the Orders in Council of 1870 and 1880, included the water and sea ice areas between the islands. During the course of his judgment, Mr Justice Morrow quoted from the statement made by Pearson in 1946 to the effect that Canada claimed sovereignty between the boundary meridians up to the Pole. However, Morrow added immediately: *'It is not declarations of sovereignty that count so much as the actual day-by-day display of sovereign rights . . .'.*[182]

Arctic pollution prevention zone delimitation based partly on sector theory (1970)

In April 1970, the Canadian Parliament adopted the *Arctic Waters Pollution Prevention Act.* The Act came into force in 1972 along with

Regulations defining, by reference to the 141st meridian, or western boundary of the Canadian sector, the 'Arctic waters' to which it applied. The Regulations specifically state that 'Arctic waters' are those 'adjacent to the mainland and islands of the Canadian Arctic within the area enclosed by the 60th parallel of the north latitude, the 141st meridian of longitude and a line measured seaward from the nearest Canadian land, a distance of 100 nautical miles . . .'.[183]

It might be significant to note that an attempt was made at the time of this enactment to include a saving clause covering the sector theory. The proposed amendment, presented by Mr Baldwin, the Progressive Conservative House leader, read:

> Nothing in this Act shall in any way be construed to be inconsistent with Canada's rightful claim of sovereignty in and over *water, ice and land areas* of the Arctic regions between the degrees of longitude 60 and longitude 141.[184]

The President of the Privy Council, Mr Donald Macdonald, intervened against the amendment on a procedural point, stating:

> The basic point is that the bill does not deal with the question of sovereignty in any sense, but deals with pollution . . . It is purely a question of pollution control jurisdiction. The bill does not say that Canada claims sovereignty over an area of the high seas a hundred miles beyond Canadian territorial waters . . . .[185]

He immediately added that 'the government would make it clear that, in taking a position against the acceptability of the amendment on procedural grounds, *it is taking no position . . . on the question of sovereignty*'.[186] The proposed amendment was ruled out of order.

Sector theory invoked for possible criminal jurisdiction (1970)
On July 16, 1970, the shooting of a man on Ice Island T-3, at 84°45.8' N. longitude and 106°24.4' W. latitude, some 185 nautical miles from the northernmost islands but within the so-called Canadian sector, occasioned further discussion of the sector theory. Although the United States did not hesitate to take jurisdiction, Canada thought it advisable to send a Note saying that it would not object if the ice island were treated as a ship for the purposes of the proceedings so as to facilitate the course of justice. It added, however, that 'the Canadian Government continues to reserve its position on the question of jurisdiction over the alleged offence . . .'.[187] Presumably, this reservation was based on the sector theory.

The sector theory was specifically raised as one of the arguments advanced by counsel for the accused in support of a motion to dismiss the indictment for lack of jurisdiction. Quoting from Poirier's motion of 1907 and Pearson's article of 1946, counsel argued that Canada had been claiming territorial sovereignty over the area in question under the sector theory. He did admit, however, that 'it is impossible to say that Canadian officials have spoken with one voice on this issue'.[188]

In his memorandum in opposition for dismissal, counsel for the prosecution devoted considerable attention to the sector theory and stated that the Canadian jurisdictional positions 'have been inconsistent and varied'.[189]

The trial judge based his finding of jurisdiction primarily on what he deemed to be an inherent power under the sovereignty of the United States to protect its interests and citizens.[190] He did not address himself to the sector theory.

On appeal to the U.S. Court of Appeals, neither counsel raised the question of the sector theory and the Court did not mention it either. Indeed, the Court of Appeals, which ordered a new trial because of an omission in the judge's charge to the jury, did not decide on the matter of jurisdiction. It simply stated that it was unable to decide the 'jurisdictional content' and went on to deal with the other grounds of appeal.[191]

Electoral boundaries based on sector theory (1976)
In 1975, the *Electoral Boundary Readjustment Act,* was amended, providing for the Northwest Territories to be divided into two redefined federal electoral districts. In 1976, the No. 1 'Western Arctic' District was described in the Canada Gazette as including the District of Mackenzie and a triangular strip north of the Yukon between the 141st and the 134th meridians of longitude, and the No. 2 District, called 'Nunatsiaq', was described as covering the rest of the Northwest Territories right up to the North Pole along the 60th meridian.[192] The newly defined districts were shown on a map entitled 'Federal Electoral Districts – 1976', published by the Department of Energy, Mines and Resources in 1976. In other words, the sector theory was used in the drawing of the boundaries of the two federal electoral districts of the Northwest Territories.

Game management boundaries based on sector theory (1976)
Hunting and trapping game regulations, made in 1976 pursuant to the *Northwest Territories Game Ordinance* of 1974, describe an area called 'Game Management Zone No. 31' as having for northern boundaries the 141st and 60th meridians of longitude right up to the Pole.[193] Such description is obviously based on the sector theory or on an interpretation of 'territories' in the Game Ordinance which coincides with that theory.

The intent is clear enough, but it is highly questionable if the term 'territories' may be given such an extensive interpretation. The Northwest Territories Act defines 'territories' as comprising 'all that part of Canada north of the 60th parallel of north latitude, except portions thereof that are within the Yukon Territory, the Province of Quebec or the Province of Newfoundland . . .'.[194] It is doubtful that this can be interpreted to include more than the land and islands transferred to Canada by the British Order in Council of 1880, plus whatever lands and islands were acquired by Canada afterwards along with the waters within the Canadian Arctic Archipelago over which Canada might have obtained title with the passage of time. To include the waters north of the Archipelago between the 141st and 60th meridians as falling within the definition of 'territories' is simply unwarranted and constitutes a reliance on the sector theory.

Fishing zone delimitation based partly on sector theory (1977)
In 1977, when Canada established an exclusive fishing zone of 200 miles in the Arctic, it described the westerly limit as being at 141° W longitude and the easterly one at 59°51'57" W longitude. Those are the meridians traditionally describing the so-called Canadian sector, except that here 59°51'57" is used instead of simply 60°.[195]

Aboriginal rights delimitation based partly on sector theory (1984)
In the Inuvialuit Final Agreement signed on June 5, 1984 with the Committee for Original Peoples' Entitlement, the 141st meridian of longitude was used to indicate the western boundary of the 'settlement region' (see Figure 8).[196] In addition to granting title over land areas traditionally used and occupied by six communities of the western Arctic, the Canadian Government purported to grant certain rights in a considerable area of the Beaufort Sea extending along the 141st meridian up to the 80th parallel of latitude. These include the exclusive right to harvest certain species of wildlife such as the polar bear and the preferential right to harvest other species of wildlife as well as marine mammals and fish.

Other possible uses of the sector theory by Canada
The above review does not purport to be exhaustive. Indeed there are a number of other, mostly indirect uses of the sector theory which may be uncovered.[197] Quite often it is impossible to determine if the use of one or both of the meridians forming the so-called Canadian sector was made for convenience of description only or with intent of ascribing an inherent legal value to it.

Fig. 8. Inuvialuit settlement region, 1984. *Source*: Dept Indian Affairs and
Northern Development, *Communiqué* No.1–8412, 5 June 1984.

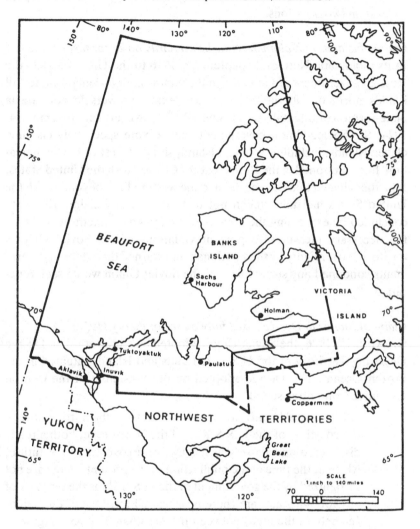

State practice of the Soviet Union
The Soviet Union has used the sector theory, in whole or in part, at least on two occasions.

Claim of islands in diplomatic Note based partly on sector theory (1924)
In its diplomatic Note of 20 September, 1916 to the United States, the Russian government did not rely on the sector theory when it claimed all islands north of Siberia as being 'une extension vers le nord de la plate-forme continentale de la Sibérie'.[198] However, on November 4, 1924, the Soviet Union government sent a Note specifically claiming eleven islands, including Wrangel Island, situated west of the 'demarcation' line mentioned in the 1867 Treaty of Cession with the United States. It further alleged that the line defined the western limit beyond which the United States had undertaken not to formulate any claim. The Note concluded by expressing the hope that foreign governments would take the necessary measures to prevent violations of the Soviet Union's territorial sovereignty by their nationals and warned that, if those governments condoned any such violation, the Soviet Union would seek reparation.[199]

Claim of islands in Decree based fully on sector theory (1926)
On April 15, 1926, the Soviet Union adopted a Decree entitled 'Decree Declaring Territory of the U.S.S.R. Lands and Islands Situated in the Frozen Ocean'. The Decree adopted by the Presidium of the Central Executive Committee stated:

> All *lands and islands* situated in the Arctic to the north, between the coastline of the U.S.S.R. and the North Pole, both already discovered and those which may be discovered in the future, which at the time of the publication of the present decree are not recognized by the government of the U.S.S.R. as the territory of any foreign state, are (hereby) declared territory of the union, (namely in the area) between the meridian 32°4'35" longitude, East of Greenwich . . . and the meridian 168°49'30" longitude, West of Greenwich . . . .[200] (see Figure 9).

It will be noted that the Decree limited itself to a claim over lands and islands and, in spite of a wider interpretation by some of the Soviet jurists, there is no definite evidence that the Soviet government has ever laid claim to the waters of the sector. Indeed, after a review of the relevant literature in the original Russian language, Professor W. E. Butler's conclusion is that 'there is nothing in subsequent Soviet legislation or

Fig. 9. Sector and jurisdictional lines in the Arctic. *Source: Polar Regions Atlas*, GC78–10040, 1978, 43.

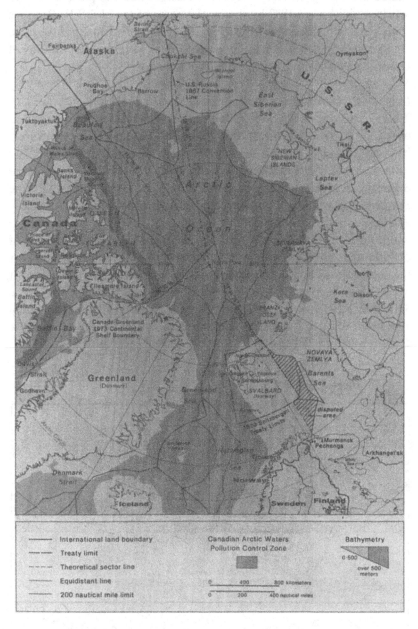

practice to suggest that the geographic coordinates laid down in the 1926 decree are a state frontier'.[201] Professor Butler refers in particular to an article, published in 1932 by the legal adviser of the U.S.S.R. People's Commissariat of Foreign Affairs, where it is made clear that the 1926 Decree merely defined the geographic limits within which the Soviet Union claimed sovereignty over lands and islands.[202]

Other possible uses of sector theory by the Soviet Union

The 1924 Note and 1928 Decree would seem to indicate the only occasions on which the Soviet Union officially relied on the sector theory. However, it is very possible that there were other occasions not so well known. For instance, it has been reported that, in its negotiations with Norway for the delimitation of the continental shelf off the Svalbard Archipelago, the Soviet Union has relied on the sector theory (or at least on sector line 32°4'45") as a special circumstance.[203]

Presuming the above use to be for the purpose of continental shelf delimitation, the Soviet Union has certainly not gone so far as to consider the sector lines as real boundaries in the Arctic Ocean. Otherwise, it would have been infringing upon the sovereignty of the other four Arctic States since 1937, when it began its use of ice islands as scientific drifting stations. Those ice stations have criss-crossed virtually the entire Arctic Ocean, wandering in all of the so-called sectors (see Figure 10).[204] This is

Fig. 10. Drifting ice stations of the Soviet Union (1937–73). *Source: Polar Geography*, 1976–7, 37.

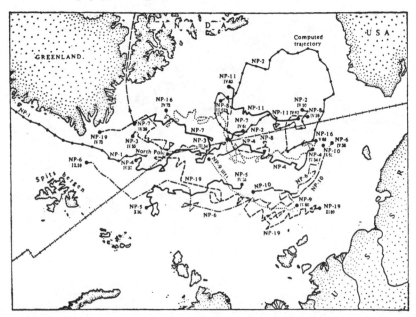

particularly so since 1972, when the Soviet Union began to maintain at least two and often three such stations at a time. It has now reached number 27 and, presuming this station continues to follow the present general direction, it should come within the Canadian sector lines eventually.

State practice of Norway

Norway has protested categorically against the use of the sector theory in the Arctic. In 1930, the Norwegian Prime Minister made it a condition precedent for Norway's recognition of Canadian sovereignty over the Sverdrup Islands that such recognition would not imply any adherence to the sector theory. And, indeed, the exchange of Notes which followed makes this quite clear. Norway's Note of August 8, 1930 reads as follows:

> At the same time, my government is anxious to emphasize that their recognizance of the sovereignty of His Britannic Majesty over these islands is *in no way based on any sanction whatever in what is named the 'sector principle'.*[205]

Norway's attitude toward the sector theory has not changed and, in its present negotiations with the Soviet Union for the delimitation of the continental shelf between Svalbard and Franz Josef Land, it has consistently refused to accept a division line approximating the meridian of $32°4'35''$ which is one of the sector lines mentioned in the 1926 Decree.[206]

Another indication of Norway's opposition to the sector theory is the claim which it made to East Greenland, resulting in a case before the Permanent Court of International Justice in 1933. During the proceedings before the Court, Norway opposed very strongly any semblance of reliance by Denmark on the doctrine of contiguity, which is at the basis of the sector theory.[207]

State practice of Denmark

The attitude of Denmark toward the sector theory might be qualified as being one of non-reliance and implied opposition. If Denmark had wished to rely on the sector theory in the 1920s, it could have claimed all of the northeast part of Ellesmere Island, situated east of the meridian of 75° longitude west, north of the 78th parallel, which includes the western projection of Greenland and over which Danish sovereignty was not contested. Instead, Denmark expressed doubt as to Canada's sovereignty on the whole of Ellesmere Island in a Note dated April 26, 1920 in which it quoted, with approval, from a report by its explorer Rasmussen, that Ellesmere Island was 'no man's land'.[208] Denmark's claim was, of course, based on the exploration and commercial activities carried on by Rasmussen, as well as the free hunting of musk oxen by Greenland Eskimos.

As for its claim of sovereignty over East Greenland, Denmark made no mention whatever of the sector theory during the proceedings before the Court in 1933. It relied exclusively on the traditional State activities. In addition, it expressly denied any reliance on the concept of contiguity, which is at the basis of the sector theory.[209]

It should also be mentioned that Denmark does not appear to have invoked the sector theory in its negotiations with Canada for the delimitation of the continental shelf in the Lincoln Sea. The sector line or 60th meridian of longitude would be as advantageous there to Denmark as the other sector line or 141st meridian is to Canada in the Beaufort Sea.

State practice of the United States

The United States has never relied on the sector theory in the Arctic and has expressed opposition to it, either directly or indirectly, on a number of occasions. In 1924, when a group of Americans took possession of Herald Island in the name of the United States, the Department of State, in acknowledging receipt of the information, pointed out there was then no information available to show that any Russian had ever landed on that island situated within the sector of the Soviet Union.[210]

In 1925, the United States Navy co-sponsored (with the National Geographic Society) the Byrd–MacMillan Expedition to verify the possible existence of 'Crocker Land' in the Arctic Ocean north of Canada. The expedition failed to obtain an exploration permit made mandatory under the recently adopted, but much publicized, amendment to the Northwest Territories Act. In the same year, the United States manifested its disregard for the sector theory by refusing to relinquish its claim to Wrangel Island situated north of Siberia. Even after both the British and Canadian governments had abandoned any claim to Wrangel Island, the State Department correspondence reveals that the United States had not abandoned its claim.[211]

In 1929, very strong opposition was expressed by the Secretary of the Navy against the suggestion of partitioning the Arctic region into five national sectors. The note from the Secretary of the Navy Adams reads as follows:

> When a private citizen suggested to president Hoover that the United States should take the initiative in bringing about an international arrangement for the partitioning of the Arctic

region into national sectors of five contiguous countries (United States, Canada, Denmark, Norway, and Russia), the proposal was called to the attention of the Navy Department, which stated that the course of action proposed –

(a) Is an effort arbitrarily to divide up a large part of the world's area amongst several countries;
(b) Contains no justification for claiming sovereignty over large areas of the world's surface;
(c) Violates the long recognized custom of establishing sovereignty over territory by right of discovery;
(d) Is in effect a claim of sovereignty over high seas, which are universally recognized as free to all nations, and is a novel attempt to create artificially a closed sea and thereby infringe the rights of all nations to the free use of this area.
I, therefore, consider that this government should not enter into any such agreement as proposed.[212]

This position with respect to the freedom of the seas being applicable in the Arctic, regardless of the sector theory, has remained the same throughout the years. The U.S. Navy has traversed frequently the various sectors of the Arctic Ocean and peripheral seas by aircraft, icebreakers and submarines.

Like the Soviet Union, the United States has operated a number of scientific stations on ice islands without regard for any sector line. In 1952, when the United States first occupied ice island T-3, located at that time within the Canadian sector, it did not request Canada's permission. It merely asked Canada to use the joint weather station at Alert as a staging point for the landing operations. Once installed on the ice island, the United States exercised exclusive jurisdiction and did not hesitate to take the necessary steps when a homicide was committed on that ice island in 1970. It exercised jurisdiction in spite of the fact that the ice island was back in the so-called Canadian sector, at approximately 185 miles north of the Canadian Arctic Archipelago. In 1970 also, the President of the United States is reported to have indicated, in a statement to Congress on February 18th, that the United States had more interest in regarding the Arctic frozen waters as high seas than as land subject to territorial sovereignty.[213]

The above activities of the United States are in conformity with an opinion of the Legal Adviser's office in 1959 that 'the United States has not recognized the so-called "Sector Principle" as a valid principle for claiming jurisdiction'.[214]

4.3.   State practice in the Antarctic

There are presently 32 States Parties to the 1959 Antarctic Treaty.[215] Of those States, 12 are original Parties, seven of which made claims to specific sectors of the continent before the conclusion of the treaty. In the order in which they formally made their claims, those States are: United Kingdom, New Zealand, Australia, France, Norway, Chile and Argentina. The other five States that have not made any claim are: Belgium, Japan, South Africa, the Soviet Union and the United States. The other 20 Parties are all non-claimant States, since the treaty had already frozen all existing claims before they had become members and prohibited any new claims while the treaty is in force.

This section reviews the position of the original 12 States in relation to the sector theory: the seven claimants and the five non-claimants.

The seven claimant States

Six of the seven claimant States have described their claim in sector form, although they have not necessarily based their claim on the sector theory and may have used meridians of longitude only as a convenient method of delimiting the extent of their claim. Norway has been the exception, and has refused to project its boundary lines as far as the South Pole (see Figure 11).

United Kingdom (1908, 1917 and 1962)

In 1908, Great Britain adopted two measures relating to the dependencies of the Falkland Islands in the Antarctic, measures which might be considered as being indirectly related to the sector theory. On July 21, letters patent made South Georgia, the South Shetland Islands, the South Sandwich Islands, the South Orkney Islands and Graham Land all dependencies of the Falkland Islands. On December 22, an Ordinance defined the expression 'dependencies' as meaning 'the groups of islands known as South Georgia, the South Orkneys, the South Shetlands, and the Sandwich Islands and the territory known as Graham's Land, situated in the South Atlantic Ocean to the south of the 50th parallel of south latitude, and lying between the 20th and the 80th degrees of west longitude'.[216]

On March 28, 1917, Great Britain issued letters patent, containing a new definition of the Falkland Islands Dependencies, so as to include 'all *islands and territories* whatsoever between the 20th degree of west longitude and the 50th parallel of south latitude; and all *islands and territories* whatsoever between the 50th degree of west longitude and the 80th degree of west longitude which are situated south of the 58th parallel of south latitude'.[217] In 1962, however, a new Order in Council separated those islands and territories of the dependencies south of the 60th parallel and named them the 'British Antarctic Territory', leaving South Georgia and the South Sandwich Islands as the Falkland Islands' Dependencies.[218] This insulated the British claim in Antarctica from the possible negative effect of any agreement with Argentina relating to the Falkland Islands.

Fig. 11. Sector lines of territorial claims in the Antarctic. *Source: Polar Regions Atlas*, GC78–10040, 1978, 43.

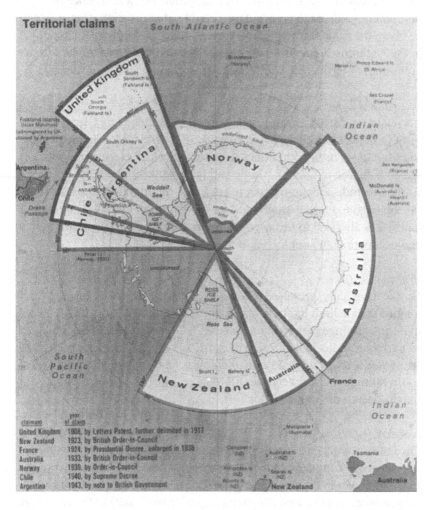

New Zealand (1923)

The New Zealand claim, known as the Ross Dependency, was officially made by Great Britain on behalf of New Zealand by the adoption of an Order in Council on July 30, 1923. The Order stated that His Majesty's Dominions in the Antarctic Seas comprised 'all of the *islands and territories* between the 160th degree of east longitude and 150th degree of west longitude, which are situated south of the 60th degree of south latitude'.[219]

Since the sector described encompasses the largest ice shelf in the world, the Ross Ice Shelf, about the size of France, it is possible that the claim might now be interpreted as including the ice shelf. The assimilation of ice shelves to land is in accord with the general view that, since they are permanently affixed to land and are capable of occupation, they ought to be given the same status as land. The assimilation seems to be made also in the Antarctic Treaty itself which specifies that the treaty applies 'to the area south of 60 degrees South Latitude, including all ice shelves'.[220]

Australia (1933)

The claim by Australia was formally made on its behalf by Great Britain, in an Order in Council dated February 7, 1933. The Order claimed 'all the *islands and territories* other than Adelie Land which are situated south of the 60th degree of South Latitude and lying between the 160th degree of East Longitude and the 45th degree of East Longitude'.[221] The claim is by far the most extensive of all but, like most of the others, it is limited to lands. It is also said to be based on the traditional principles of discovery and occupation and not the sector theory.[222]

France (1938)

France originally proposed to use the 136th and 147th degrees east longitude to delimit its claim to a sector called 'Adelie land' but decided, after an exchange of correspondence with Great Britain, to confine itself to narrower limits in its Decree of April 1, 1938.[223] The Decree claimed 'the *islands and territories* situated south of the 60-degree parallel of south latitude and between the 136-degree and 142-degree meridians of longitude east of Greenwich'.[224]

Norway (1939)

On January 14, 1939, Norway adopted a Royal Decree bringing part of the Antarctic continent under its sovereignty. It stated: 'That part of the *mainland coast* in the Antarctic which extends from the Falkland Islands Dependencies in the west (the boundary of Coats Land) to the boundary of the Australian Antarctic Dependency in the east (longitude 45°E) *together with the land situated inside this shore and the sea adjacent thereto* is brought under Norwegian sovereignty'.[225] It will be noted that this claim is one limited to the coast and to an undefined hinterland. It was stated at the time that 'the region occupied is not claimed as a sector . . . . It is the coast together with the land lying inside (hinterland) and . . . the appropriate sea territory'.[226]

Consistent with its position on the sector theory, Norway has claimed sovereignty over Peter I Island, situated close to the meridian of 90° west, outside its sector lines and on the edge of Bellingshausen Sea. Norway's limited claim in 1939 was also in conformity with the communication previously made to the State Department. The Memorandum of the State Department reads: 'The Minister for Foreign Affairs told me the other day that Norway has no intention of annexing territory charted by the *Norvegia*, but that it would object to applying the sector principle to the south polar regions and that freedom of the seas would be claimed.'[227]

Chile (1940)

On November 6, 1940, Chile adopted a decree making the following claim:

> All *lands, islands, inlets, reefs of rocks, glaciers (pack-ice),* already known, or to be discovered, and their respective terri-torial waters, in the sector between longitudes 53 and 90 West, constitute the Chilean Antarctic or Chilean Antarctic terri-tory.[228]

Presumably, the term 'pack-ice' which appears in the Spanish text, in brackets after the word 'glaciares', is meant to refer to that part of the Ronne Ice Shelf within the sector and not to sea ice. As stated earlier with respect to the Ross Ice Shelf, the assimilation of an ice shelf to land would seem to be legally acceptable. Another reason for presuming that 'glaciares' is meant to refer to ice shelf and not to sea ice is that Chile bases its claim on the old principle of *uti possidetis juris*.

It will be noted that the eastern limit of longitude 53° West includes the Antarctic Peninsula and the South Shetland Islands, both of which are also claimed by the United Kingdom. Indeed, the Chilean claim overlaps some 27 degrees with that of the United Kingdom.

Argentina (1943)

Argentina has not adopted any decree incorporating its claim but it took a number of official steps indicating a formal claim to a specific sector. In January 1942, the commander of the Argentine naval ship *Primero de*

Mayo took possession of Deception Island in the South Shetlands and deposited an act of possession claiming all territory bounded by longitudes 25° and 68°34' West and latitude 60° South.[229]

The above was followed by a Memorandum to the United Kingdom on 15 February 1943 which read as follows:

> The Argentine Government reaffirms on this occasion its sovereign rights over all Antarctic *lands and dependencies* situated south of parallel 60° south latitude, and between meridians 25° and 68°34' west longitude.[230]

The limits of the sector claimed by Argentina have remained the same, although it published a map in December 1946 showing the limits to be longitudes 25° and 74° West. [231] These limits include all of Palmer Island and most of Alexander Island, but the map was not followed by any diplomatic Note or Decree officially enlarging the claim. The Argentine sector lies wholly within that claimed by the United Kingdom and jointly within that claimed by Chile.

The five non-claimant States
United States
The United States has expressed definite opposition to sector sovereignty claims in the Antarctic. An example of this opposition is found in a Note sent to Chile in 1959, as follows:

> In the note of the Embassy of Chile reference is made to the 'Chilean Antarctic sector'. As is well-known, the Government of the United States has never recognized the claims of sovereignty which have been asserted by seven countries in Antarctica, but has consistently reserved its basic historic rights in that region.[232]

Indeed, the United States has expressly communicated its non-recognition of sovereignty claims to other claimant States and the reservation as to its own rights, in a number of diplomatic Notes. For instance, it has done this in a Note to Australia on August 13, 1955[233] and in another to the United Kingdom on March 23, 1962.[234]

At the time it proposed the conference on Antarctica in 1958, the United States had also been careful to reserve 'all of the rights of the United States with respect to the Antarctic region, including the right to assert a territorial claim or claims'.[235] This Note, proposing the conference, was addressed to the foreign ministers of Argentina, Australia, Belgium, Chile, France, Japan, New Zealand, Norway, the Union of South Africa, the U.S.S.R., and the United Kingdom.

Since it proposed the conference, the United States has concluded agreements with a number of other countries (Argentina, 1958; Australia, 1958; New Zealand, 1958; and Chile, 1960) for the operation of scientific stations. At the moment, the United States operates at least three permanent stations: one within the sector claimed by Chile, Argentina and the United Kingdom, another within the sector claimed by New Zealand and the third at the South Pole itself.

Soviet Union

The position of the Soviet Union in the Antarctic has also been one of non-recognition and non-claim. In 1950, it sent a Note to Great Britain, France, Norway, Australia, Argentina, New Zealand and the United States in which it stated that 'the Government of the U.S.S.R. cannot agree to such a question as that of the Antarctic regime being settled without its participation'.[236] The Note specified that the Antarctic continent had been discovered by Russian navigators.

In 1958, when the U.S.S.R. accepted the invitation of the United States to participate at the conference on Antarctica, it formally reiterated its position as follows:

> As to the question of territorial claims advanced by certain states in the Antarctic, the Soviet Government considers it necessary to state once again that it has not recognized and cannot recognize as legitimate any kind of separate solution for the problem of territorial possessions in (state sovereignty over) the Antarctic. In this connection it is appropriate to recall the outstanding merits of Russian explorers in discovering the Antarctic and in particular, the generally recognized fact that *it was the Russian navigators Bellingshausen and Lazarev who were the first to reach the shores of the Antarctic and circumnavigate this continent* at the beginning of the 19th century.
>
> 'The Soviet Union reserves to itself all rights based on discoveries and explorations of Russian navigators and scientists, including the right to present territorial claims in the Antarctic.'[237]

In addition to the above reservation with respect to all claimants in the Antarctic, the Soviet Union is reported to have reserved its position separately in relation to Norway in 1939.[238] At the moment, the Soviet Union is operating seven scientific stations: four within the Australian sector, one in the Norwegian sector, one in the unclaimed sector and the last one off the Antarctica Peninsula within the overlapping sectors claimed by Chile, Argentina and the United Kingdom.[239]

The other non-claimant States

The other three States, with activities of longstanding in the Antarctic and original Parties to the 1959 Treaty, are Belgium, Japan and South Africa. They have never made any claim to a sector and do not seem to have taken a public position in respect of the sectors claimed by the other States. As for Japan, it has renounced 'all claim to any right or title or interest in connection with any part of the Antarctic area'.[240]

4.4.   Appraisal of customary law as a basis for the sector theory

At the outset of this appraisal it is necessary to extract the main characteristics of the State practice already examined, particularly with reference to its content and uniformity of application. It is all the more important to determine if some form of use of the sector theory has developed into a rule of customary international law, since only two of the five Arctic States have engaged in the practice, Canada and the Soviet Union.

Canada has never adopted any law or Order in Council claiming a sector, but it has taken a number of official steps indicating reliance on the sector theory for a number of purposes and in various ways. It is, indeed, the manner in which the theory has been applied which constitutes the main difficulty in finding the necessary elements of uniformity and consistency, aside from the fact that Canada seems to have abandoned its reliance on the theory at one point for a period of about 15 years.

Canada's practice began by Captain Bernier's formal taking of possession of the whole Arctic Archipelago within the defined sector right up to the North Pole. His mandate was to annex 'territory of British possessions' and, presumably, the inscription on the bronze plaque deposited on Melville Island was not meant to include the water areas.[241] Whatever the precise intention, in 1913 Canada sent an official expedition to Wrangel Island, members of which took possession of it the next year and, in 1922, the Prime Minister affirmed Canada's claim to the island.

In 1925, Canada developed doubts as to the effectiveness of its claim to some of the northernmost islands and feared the discovery of new ones within its sector by foreign explorers. It therefore invoked the sector theory again and claimed sovereignty right up to the North Pole, but with clear reference to land. The following year, it created the Arctic Islands Preserve extending to the Pole and in 1959, adopted new game regulations to apply within the Preserve.

Between 1938 and 1954, one Minister and the Prime Minister made statements in the House of Commons, both claiming sovereignty rights up to the Pole, but without any specific reference to land or water. Then, from 1955 to 1965, various statements by Ministers and two Prime Ministers invoked the sector theory, some with reference to land only and one without any specific reference. In 1965, the western sector line was used as a boundary for the issuance of offshore oil and gas permits and, in 1969, the Prime Minister denied the application of the sector theory to water areas and limited it to the continental shelf. Since 1970, one or both of the sector lines have been used to delimit various zones of jurisdiction, including the following: pollution prevention, electoral district delimitation, fishing and game management

No consistent pattern or position emerges from Canada's use of the sector theory. Indeed, the purpose and manner of the use, as well as its non-use for a period of time, seemed to have been dictated by the needs of the moment.

In addition to the above, Canada has been publishing maps since 1897, showing Canada's boundaries following the sector lines. Although those boundary lines did not extend to the North Pole on the 1897 map, they did so on the 1904 map and on virtually all maps published since by the Department of the Interior and later by the Department of Energy, Mines and Resources. Since 1952, the boundary lines extending to the North Pole have been shown on maps of Canada for general circulation and a great number of other maps and charts.

Maps have often been taken into account as part of the subsequent conduct of the Parties as an aid for the interpretation of a treaty, where the exact position of the boundary is at issue. But here there is no question of a treaty being involved to determine the eastern line of the sector. As for the western line, the analysis already made of the boundary treaties of 1825 and 1867 has revealed that the meridians of longitude were used only to delimit the territorial possessions of the Parties. More precisely, the boundary between Alaska and the Yukon territory, following the 141st meridian, stops at a point on the coast at about 69°39' latitude north, indicated as 'Demarcation Point' on British Admiralty charts as well as Canadian charts produced by the Hydrographic Service.[242]

The above was the boundary followed by Canada when describing the Yukon Judicial District in the Proclamation of 1897 included in the Yukon Territory Act of 1898. The relevant part of the Proclamation reads:

> Beginning at the intersection of the 141st meridian of west longitude from Greenwich with a point on the coast of the Arctic Sea, which is approximate north latitude, 69°39', and named on the Admiralty charts 'Demarcation Point'; then due south thereafter to run due north to the Arctic Ocean, or the westernmost

channel of the Mackenzie Delta, and along that channel to the Arctic Ocean; thence north-westerly following the windings of the Arctic Coast (termination of the mainland of the Continent), including Herschel Island, and all other islands which may be situated within three (3) geographic miles, to the place of beginning.[243]

Considering the above, and regardless of the probative value of maps and charts in certain circumstances, it would be of little if any assistance for Canada to invoke its maps in support of its own claim. This is particularly so since the other Arctic States have not reproduced Canada's sector lines on their own maps.[244]

As for the Soviet Union, it adopted a formal Decree in 1926, following two diplomatic Notes of 1916 and 1924, claiming all lands and islands within the sector lines defined in the Decree. Although a number of its jurists have interpreted the Decree as including a claim to the water area, the Soviet government has never done so, and the long use of all parts of the Arctic Ocean by its drifting scientific stations militate against any such claim. It is possible, however, that the Soviet Union has relied on a sector line in its negotiations with Norway for the delimitation of the continental shelf. In the Antarctic, the Soviet Union is not making any claim and has refused to recognize any of the sector claims.

The practice of the other three Arctic States has been quite clear and consistent: they have never relied on the sector theory. Norway strongly opposed it in the Arctic with respect to its recognition of Canada's sovereignty over the Sverdrup Islands in 1930, and it is presently objecting to its apparent use by the Soviet Union in the continental shelf negotiations. In the Antarctic, it specified merely that its claim was to the coast and the hinterland but not to the apex of the sector. As for Denmark, not only has it not relied on the theory, but it indirectly challenged Canada's sector claim in its Note of 1920 when it stated that it considered Ellesmere Island, in good part outside a possible Danish sector, as a 'no-man's land'.

The United States has never relied on the sector theory and has often expressed opposition to it, particularly as a basis for claiming sovereignty over part of the Arctic Ocean. It has sent numerous submarines in that ocean regardless of sector claims and, like the Soviet Union, has operated a number of scientific drifting stations which have criss-crossed the various sectors. In the Antarctic, it is not making any claim and is not recognizing any.

Has the State practice of the five Arctic States just reviewed been sufficiently uniform and received the necessary degree of acceptance to have resulted in a rule of regional customary law binding on all Arctic

States? It seems obvious that the answer to both parts of that question has to be in the negative. Firstly, the practice of the two proponents has not been uniform or consistent, since the theory has been invoked for a number of different purposes depending on the need of the moment. Secondly, the other three Arctic States have not relied on the theory and two have strongly opposed it.

The question remains whether State practice in the Antarctic, when added to State practice in the Arctic, is such as to give additional validity to the theory and possibly promote it to a special rule of customary law for the acquisition of sovereignty in polar regions. An examination of the Antarctic claims reveals two important characteristics. The first is that no claim is expressly described as extending to the apex of the sector at the South Pole, and the second is that all but the Chilean claim are limited to lands and islands. The claim of Chile includes the shelf ice and, as explained previously, the assimilation of the shelf to land is generally accepted, even in the Antarctic Treaty, because of the permanency of its attachment to land. In these circumstances and presuming for the moment that Antarctic sector claims are generally recognized (which is certainly not the case), the most that could be said is that the sector theory could serve as a basis for a claim to sovereignty over land areas, with or without shelf ice.

The above examination of State practice in both the Arctic and the Antarctic leads to the conclusion that *the sector theory has not developed as a principle of customary law,* neither general nor regional, and cannot serve as a root of title for the acquisition of sovereignty, particularly not to areas of the sea. In the Arctic, the practice has been followed only by Canada and the Soviet Union, and it is not clear if their intention was to rely on the theory as a legal basis for their claim or simply as a means of describing its geographic extent. Neither has it been always clear, especially in the case of Canada, if the claim is related to land, sea, continental shelf or to a combination of the three. The other three Arctic States have never relied on the sector theory for any purpose, and the United States and Norway have consistently opposed its use. Norway has even refused to use a full sector to describe its claim in the Antarctic, for fear that its use might be misunderstood. In addition, both the United States and the Soviet Union have objected to all sector claims in the Antarctic and have made no claims themselves. In these circumstances, there could never have arisen a general or a regional practice accepted as law.

Notes to Part 1

1. See Papal Bull *Inter Caetera*, reproduced (English translation) in W. M. Bush, *Antarctica and International Law*, Vol. 1, 702 pages, at 532 (1982).
2. *Ibid.*, at 533.
3. Reproduction in *European Treaties Bearing on the History of the United States and its Dependencies*, Vol. I, at 146 (1917).
4. See Order in Council, 23 June 1870, reproduced in *Statutes of Canada*, 1872, at lxiii.
5. See *Map of North America*, ordered printed by the Canadian House of Commons on 12 July 1850 and showing in green the 'territories deemed by the Hudson's Bay Company in virtue of the Charter granted to them by King Charles the Second'.
6. *Can. Sen. Debates*, 3 May 1978, at 903; emphasis added.
7. See Order in Council, 31 July 1880, *Canada Gazette*, Vol. XIV, No. 15. For a study of this transfer see Gordon W. Smith, 'The Transfer of Arctic Territories from Great Britain to Canada in 1880 . . .', 14 *ARCTIC* 53–7 (1961).
8. *Canada Gazette*, 14 May 1898, at 2613.
9. This map was submitted with Order in Council P.C. 3388 of 18 Dec. 1897. See Public Archives of Canada, Call No. R.G. 2, Series 1, Vol. 752.
10. Andrew Taylor, *Geographical Discovery and Exploration of the Queen Elizabeth Islands*, 172 pages, at 112 (1964).
11. *Can. Sen. Debates*, 20 Feb. 1907, at 266; emphasis added.
12. *Ibid.*, at 271; emphasis added.
13. *Ibid.*, emphasis added.
14. *Ibid.*, at 274.
15. *Ibid.*
16. *Ibid.*
17. 'Convention entre Grande-Bretagne et Russie, concernant les limites de leurs possessions respectives sur la côte Nord-Ouest de l'Amérique . . .', 16/28 janvier 1825, *Recueil de Martens*, Nouvelle série, Vol. II, at 427–8. The Convention was concluded in French only. For both French and English versions, see *Consolidated Treaty Series*, Vol. 75, at 96 and reproduced in *Appendix I*.
18. Art. I, Convention ceding Alaska between Russia and the United States, *Consolidated Treaty Series*, Vol. 134, at 332, reproduced in *Appendix II*.
19. D. H. Miller, 'Political Rights in the Arctic', 4 *Foreign Affairs*, 47-60, at 59 (1925); emphasis added.
20. *Ibid.*
21. *Ibid.*, emphasis added.

22. Before him, Leonid Breitfuss had suggested a division of the Arctic into five sectors; see Leonid Breitfuss, 'Territorial Division of the Arctic', 8 *Dalhousie Review* 456–70, at 467–8 (1929). He did not use the boundary treaties as a basis, so his suggestion is not discussed here.
23. W. L. Lakhtine, 'Rights over the Arctic', 24 *A.J.I.L.* 703–17, at 715 (1930). This is a translation of the French version 'La voie aérienne arctique et l'état juridique des territoires polaires septentrionaux', *Revue de droit aérien* 532–56 (1929).
24. *Ibid.*, at 715–6.
25. *Ibid.*, footnote at 716 (1930).
26. *Ibid.*, at 716.
27. *Ibid.*
28. See 1920 Spitzbergen Treaty Limits on Figure 9 in Chapter IV, *infra*, as per Art. I of the Paris Convention.
29. *Supra* note 23, at 716.
30. *Eastern Greenland Case* 1933 P.C.I.J., Series A/B, at 74, French text.
31. *Supra* note 23, at 716.
32. See *supra* note 17.
33. *Ibid.*
34. *Ibid.*, emphasis added. The meaning of 'ligne de démarcation' will be examined later.
35. For an excellent study of this award, see C. B. Bourne and D. M. McRae,' Maritime Jurisdiction in the Dixon Entrance: The Alaska Boundary Re-examined', 14 *C.Y.I.L.* 175–223 (1976).
36. See *supra* note 1; emphasis added.
37. *Grand Dictionnaire Universel du XIXe siècle*, Vol. 9, at 1124 (1872).
38. *Affaire du Monastère de Saint-Naoum* (1924) P.C.I.J., Series B, Vol. 1, No. 9, at 17.
39. *Ibid.*, at 20.
40. *Ibid.*, Series C, No. 5–11, at 34; emphasis added.
41. See *supra* note 37, Vol. 8 at 1286.
42. See de Martens, *Nouveau Recueil de Traités*, Vol. 2, at 682.
43. See A. G. de Lapradelle, *La Frontière*, 368 pages, at 11 (1928).
44. The only possible exception to this relates to the beginning of the line describing the southern limit of the panhandle, across the water area between Prince of Wales Island at the entrance of Portland Canal (known as the A–B line since the 1903 arbitral award), which could conceivably be considered a true boundary line rather than a mere line of allocation.
45. Art. III, *supra* note 17; emphasis added.
46. See *Dictionnaire de la Terminologie du Droit International* at 462 (1960).
47. *Ibid.*
48. *Ibid.*, at 596–9.
49. See *supra* note 17.
50. See *supra* note 18.
51. Art. I, *ibid.*
52. See Bering Sea Seal Fishery Arbitration, 85 *British and Foreign State Papers* 1158. The arbitral tribunal in effect adopted the British argument that 'the line drawn through Behring Sea between Russian and United States possessions was thus intended and regarding (*sic*) merely as a ready and definite mode of indicating which of the numerous islands in a partially explored sea should belong to either Power'. See Parliamentary Paper, United States No. 1, 1893, quoted in W. M. Bush, *Antarctica and International Law*, Vol. II, at 127 (1982).
53. See 'Coordinate Positions for the Plot of U.S.–Russia Convention of 1867',

International Boundary Study, No. 14, October 1, 1965, published by U.S. Department of State.

54. V. Stefansson, *The Adventure of Wrangel Island*, at 20 (1925).
55. Art. I, *supra* note 18; emphasis added.
56. Reproduced in C. B . Bourne and D. M. McRae, *supra* note 35, at 199.
57. See *supra* note 53.
58. *Ibid.*, at 3; emphasis added.
59. S. W. Boggs, 'Delimitation of Seaward Areas Under National Jurisdiction', 45 *A.J.I.L.*, at 240, footnote 2 (1951)
60. 1898 Stat. of Can., Chap. 6.
61. See *Island of Bulama Case* (1870) in Moore, *History and Digest of International Arbitrations*, Vol. II, 1909–1922, at 1921.
62. Moore, *Digest of International Law*, Vol. I, at 264 (1906).
63. For the diplomatic correspondence on this controversy, see Miller, *Treaties and Other International Acts of the U.S.A.*, Vol. 5 (1846–52), 3–101 (1935).
64. *Supra* note 62, at 265.
65. *Ibid.*, at 266.
66. *Ibid.*, at 267.
67. *Can. Sen. Debates*, 1906–7, at 270.
68. Note reproduced in W. L. Lakhtine, *Rights Over the Arctic* (in Russian), Moscow, at 44 (1928).
69. *Ibid.*
70. *Documents on Canadian External Relations*, Vol. 3, at 569 (1919–25).
71. *Can. H.C. Debates*, 1924, at 1111.
72. See dispatch from British Minister of 23 Dec. 1929, quoted in W. M. Bush, *Antarctica and International Law*, Vol. II, at 130 (1982).
73. Declaration reproduced in Whiteman, *Digest of International Law*, Vol. 2, at 1257 (1963).
74. See De Lapradelle and Politis, *Recueil des arbitrages internationaux* 404–31, at 413 (1932).
75. Moore, *History and Digest of International Arbitrations*, Vol. II, 1909–1922, at 1921.
76. *Ibid.*
77. Moore, *supra* note 75, Vol. IV, 3354–9, at 3357.
78. United Nations, *Rep. Int'l Arbitral Awards*, Vol. XI, 21–3, at 23.
79. *Ibid.*, at 21–2.
80. *Supra*, note 78, Vol. II, at 854.
81. *Ibid.*
82. *Ibid.*, at 855.
83. *Ibid.*
84. *Ibid.*, at 869.
85. [1933] P.C.I.J., Series C, No. 63, at 747.
86. *Ibid.*, No. 66, at 3239–45.
87. *Ibid.*, at 3270.
88. See [1953] I.C.J. Memorials, Vol. I, at 239.
89. [1953] I.C.J. Rep., at 58–9.
90. [1975] I.C.J. Rep., at 35, para. 92.
91. *Ibid.*
92. *I.L.C. Yearbook*, Vol. II, at 298 (1956).
93. [1969] I.C.J. Rep., at 30.
94. *Ibid.*, at 30–1. For completeness, it should be mentioned that two dissenting judges, Morelli and Tanaka, expressed a favourable view on the concept of contiguity; for their respective opinions, see *id.*, at 201 and 180. On the other

hand, Judge Ammoun, in a separate concurring opinion, rejected completely the argument based on contiguity; see *id.*, at 115.

95. See Decision of 30 June 1977, typewritten text of 236 pages, at 90, para. 81.
96. *Ibid.*
97. *Ibid.*
98. [1982] I.C.J. Rep., at 61, para. 73.
99. *Ibid.*
100. See judgment dated 12 October 1984, paragraphs 103 and 104; emphasis in text. The 'distance principle' was recognized, by the International Court in the *Malta/Libya Case*, to entitle a coastal State to continental shelf rights, but strictly within the new concept of the 200-mile exclusive economic zone. See judgment dated 3 June 1985, paragraphs 39 and 43.
101. *Supra*, note 93, at 116; emphasis added.
102. P. Fauchille, *Traité de Droit international public*, Vol. I, at 733 (1925). For a review of the opinions of writers, see *id.*, at 752 and the Counter-Memorial of Norway in the *Eastern Greenland Case* [1933] P.C.I.J., Series C, No. 62, at 457 ff.
103. D. H. Miller, 'Political Rights in the Arctic', 4 *Foreign Affairs* 47–60, at 56 (1925).
104. For a discussion of this agreement, see Quincy Wright, 'Territorial Propinquity', 12 *A.J.I.L.* 519–61, at 519 (1918).
105. *Supra*, note 103, at 56–7.
106. M. F. Lindley, *The Acquisition and Government of Backward Territory in International Law*, at 235 (1926).
107. *Ibid.*, at 230.
108. W. L. Lakhtine, 'Rights over the Arctic', *A.J.I.L.* 703–717 (1930).
109. See G. Smedal, *De l'acquisition de souveraineté sur les territoires polaires*, at 52 and 92 (1932).
110. See K. Strupp, 'Les conditions d'une occupation valable aux diverses époques importantes au Groenland Oriental', [1933] P.C.I.J., Series C, No. 64, at 1905–6.
111. C. C. Hyde, 'Acquisition of Sovereignty Over Polar Areas', 19 *Iowa L. Rev.* 286–94, at 289 and 293–4 (1934).
112. F.A.F. von der Heydte, 'Discovery, Symbolic Annexation and Virtual Effectiveness in International Law', 29 *A.J.I.L.* 448–71, at 471 (1935).
113. G. Gidel, *Aspects juridiques de la lutte pour l'Antarctique*, at 36 (1948).
114. C. H. M. Waldock, 'Disputed Sovereignty on the Falkland Islands Dependencies', 25 *B.Y.I.L.* 331–53, at 341–2 (1948).
115. *Ibid.*, at 343.
116. *Ibid.*, at 345.
117. G. W. Smith, *The Historical and Legal Background of Canada's Arctic Claims*, Ph.D. dissertation, Columbia University, New York, 505 pages (1952).
118. C.M. Franklin and V. C. McClintock, 'The Territorial Claims of Nations in the Arctic: an Appraisal', 5 *Oklahoma L. Rev.* 37–48 (1952).
119. O. Svarlien, 'The Sector Principle in Law and Practice', Vol. 10 *Polar Record* 248–63, at 256 (1960).
120. F. M. Auburn, *The Ross Dependency*, 91 pages, at 24 (1972).
121. G. D. Triggs, *Australia's Sovereignty in Antarctica: the Validity of Australia's Claim at International Law*, Ph.D. dissertation, University of Melbourne Law School, at 215 (1983).
122. H. Lauterpacht, 'Sovereignty over Submarine Areas', 27 *B.Y.I.L.* 376–433, at 431 (1950).
123. See the Convention on the Law of the Sea, A/CONF. 62/122 (1982), Art. 56 (jurisdiction in the exclusive economic zone) and Art. 76 (jurisdiction over the continental shelf).

124. J. S. Reeves, 'Editorial Comment: Antarctic Sectors', 33 *A.J.I.L.* 519–21, at 521 (1939).
125. *Supra* note 114, at 338–9.
126. Art. 38, para 1(b) of the Statute of the International Court of Justice.
127. *North Sea Continental Shelf Cases* (1969) I.C.J. Rep., at 44.
128. *Asylum Case* (1950) I.C.J. Rep., at 276.
129. *Ibid.*, at 277.
130. See photograph of plaque with inscription in J. E. Bernier, *Master Mariner and Arctic Explorer*; Ottawa: Le Droit, at 128 (1939).
131. See *Débats du Sénat*, 1 Feb. 1910, 1909–1910, at 197. The expression used by Senator Poirier was 'terres et îles arctiques'.
132. *Ibid.*, 197–203.
133. See letter reproduced in V. Stefansson, *The Friendly Arctic*, at p. XXII (1944).
134. For a history of the adventure of the *Karluk*, see a fascinating account by one of the survivors, W. L. McKinlay, *Karluk*, 170 pages (1976).
135. See photograph of the Canadian Ensign being raised on July 1, 1914, in V. Stefansson, *The Adventure of Wrangel Island*; inserted after p. 60 (1925).
136. See *Report of Advisory Technical Board on the Question of Canadian Sovereignty in the Arctic*, 6 November 1920, at 2A.
137. *Ibid.*, at 3A. For a discussion of what he termed 'the supposed Danish challenge to Canada's ownership of Ellesmere Island', see Trevor Lloyd, 'Knud Rasmussen and the Arctic Island Preserve', *MUSK-OX 25*, 85-90, at 86 (1979).
138. See *Documents on Canadian External Relations*, Vol. 3 (1919–25), at 568 (1970).
139. *Ibid.*, at 568–9.
140. See letter reproduced in V. Stefansson, *The Adventure of Wrangel Island*, at 98 (1925).
141. For the text of the proclamation, see *ibid.*, at 121.
142. *Can. H. C. Debates*, at 1769 (1922).
143. *Can. H. C. Debates*, at 1110 (1924).
144. *Can. H. C. Debates*, at 3773 (1925).
145. *Ibid.*, emphasis added.
146. See D. H. Dinwoodie, 'Arctic Controversy: the 1925 Byrd-MacMillan Expedition', 53 *Canadian Historical Review* 51, at 56–8 (1972).
147. *Supra* note 143.
148. For an account of the incident, see Dinwoodie, *supra* note 146, at 62. It should be noted that, in his subsequent expeditions of 1926 and 1927–28 to Baffin Island, MacMillan did obtain the necessary licence; see Dinwoodie, *id.*, at 63–4.
149. See *supra* note 138, at 583.
150. *Canada Gazette*, 31 July 1926, at 382.
151. Attached to P.C. No. 1146, 19 July 1926. Obviously, the '65th' degree of longitude mentioned in the Memorandum should have been '60th' instead.
152. *Ibid.*
153. See P.C. 807, *Canada Gazette*, 25 May 1929, at 4018.
154. *Can. H. C. Debates*, at 3081 (1938); emphasis added.
155. *Ibid.*
156. See P.C. 6115, *Canada Gazette*, 1945.
157. *Documents on Canadian External Relations*, Vol. 12, 1946, at 1562–63; emphasis added.
158. *Ibid.*, at 1566; emphasis added.
159. *Ibid.*, at 1568.
160. 24 *Foreign Affairs*, 638–47, at 638–9; emphasis added.
161. *Ibid.*, at 641.

162. L. B. Pearson, 'Canada's Northern Horizon', 31 *Foreign Affairs* 581–91, at 586–7 (1953).
163. *Can. H. C. Debates*, at 700 (1953–4). Parts of his speech, including the one just quoted, were reproduced in the publication of the External Affairs Deparment under the heading 'Statements of Government Policy'; see *External Affairs*, Vol. VI, at 16 (1954).
164. See *Canada Yearbook*, at 1 (1954). A similar description had first appeared in the Canada Yearbook of 1934–5.
165. *Proceedings of the Special Committee on Estimates*, No. 15, at 446 (23 March 1955); emphasis added.
166. *Ibid.*
167. *Can. H. C. Debates*, at 6955 (1956); emphasis added.
168. *Ibid.*, at 6955-6.
169. *Can. H. C. Debates*, at 1993 (1957); emphasis added.
170. Ibid., emphasis added.
171. *Can. H. C. Debates*, at 1559 (1957–8).
172. *Can. H. C. Debates*, at 1963 (1958); emphasis added.
173. *Can. H. C. Debates*, at 3511 (1958).
174. *Ibid.*, at 3512; emphasis added.
175. *Ibid.*, at 3540.
176. *Ibid.*, at 3652; emphasis added.
177. *Can. H. C. Debates*, at 7657 (1964).
178. *Can. H. C. Debates*, at 6396 (1969).
179. *Ibid.*, emphasis added.
180. *Ibid.*
181. See *Regina v. Toatalik, E4–321* (1969) 71 W.W.R. 435, at 443.
182. *Ibid.*, at 439; emphasis added.
183. S.3, s-s.1, Chap. 47, 18–19 Eliz. II, 1969–70. This special definition of 'Arctic waters' was also incorporated by reference in the Ocean Dumping Control Act of 1975 in its definition of the 'sea'; see *Statutes of Canada*, 1975, c.55,s.l.
184. *Can. H. C. Debates*, at 7899 (9 June 1970); emphasis added.
185. *Ibid.*
186. *Ibid.*, emphasis added.
187. This part of the Note was read into the record by the trial judge: see *U.S.A. v. Escamilla* (No. 71–1575), Brief for Appellee, in Appendix at 18 (1971).
188. *Memorandum in Support of Motion to Dismiss Indictment for Lack of Jurisdiction, U.S.A. v. Escamilla*, Criminal No. 210–70A, at 31.
189. *Memorandum in Opposition of Motion for Dismissal for Lack of Jurisdiction, U.S.A. v. Escamilla*, Criminal No. 210–70A, at 10.
190. *U.S.A. v. Escamilla*, No. 71–1575, Brief for Appellant, at 12.
191. See *U.S.A. v. Escamilla*, No. 71–1575, judgment of the six-member Court of the Fourth Circuit sitting in banco, written by Judge Winter, at 7, 17 August 1972.
192. SI/76–76 *Canada Gazette*, Part II, Vol. 110, No. 13, 1949, at 2078–9 (14 July 1976).
193. See Northwest Territories Game Regulations No. 317–76 and Northwest Territories Game Ordinance, *N.W.T. Regulations and Ordinances*, 1974, c.G-1.
194. *Northwest Territories Act*, R.S.C. 1970, c.N-22, s.3.
195. See SOR/77–173, *Canada Gazette*, Part II, Vol. 111, No. 6, 652, at p. 653 (3 March 1977).
196. See Compensation Agreement with the Committee for Original Peoples' Entitlement (C.O.P.E.), reported in Dep't Indian Affairs and Northern Development, *Communiqué* No. 1–8359 (28 March 1984).
197. For instance, it might be possible to say that the delimitation of the aerial space

for which Canada is responsible under the International Civil Aviation Organization is partly based on the sector theory. The same might be said of Canada's Contingency Agreement with the United States relating to marine pollution arising out of drilling in the Beaufort Sea.

198. For the text of the Note in French, see Appendix in W. L. Lakhtine, *Rights over the Arctic* (in Russian), Moscow, (1928).
199. *Ibid.*
200. Hudson, *Cases on International Law*, at 220 (1951); emphasis added. See also French text in W. L. Lakhtine, *supra* note 198, at 45.
201. Wm E. Butler, *Northeast Arctic Passage*, at 77 (1978).
202. *Ibid.*, at 73.
203. K. Treavid and W. Ostreng, 'Security and Ocean Law: Norway and the Soviet Union in the Barents Sea', 4 *Ocean Development and International Law*, 343–67, at 355–6 (1977).
204. See A. F. Treshnikov *et al.*, 'Results of Oceanological Investigations by North Pole Drifting Stations' 1 *Polar Geography* 22-40, at 37 (1976–7). Note also the sector lines on the map.
205. *Canada Treaty Series*, No. 17, at 3 (1930); emphasis added.
206. See *supra* note 203, at 356–7.
207. See *supra*, Chapter III, Section II 'Contiguity in International Decisions'.
208. See *supra,* note 136.
209. See *supra*, note 207.
210. Hackworth, *Digest of International Law*, Vol. 1, at 465 (1949).
211. *Ibid.*, at 464.
212. *Ibid.*, at 463–4.
213. See *N.Y. Times*, 19 Feb. 1970, at 1, col. 8.
214. Letter of Assistant Legal Adviser M. Whiteman to Lt.-Cmdr. John C. Fry, USN, 7 January 1959, in M.S. Dep't State File 703.022/12–1758, quoted in Whiteman, *Digest of International Law*, Vol. 2, at 1268.
215. The States parties are:

Argentina*	Czechoslovakia	Italy	Rumania
Australia*	Denmark	Japan*	South Africa*
Belgium*	FRG*	The Netherlands	Soviet Union*
Brazil*	Finland	New Zealand*	Spain
Bulgaria	France*	Norway*	Sweden
Chile*	GDR	Papua New Guinea	United Kingdom*
China*	Hungary	Peru	United States*
Cuba	India*	Poland*	Uruguay*

The asterisk indicates Parties with consultative status. For a recent study on the 'Question of Antarctica', see *U.N. Doc. A/39/583* (Part I), 31 Oct. 1984; (Part II), 29 Oct. 1984; (Part II), 2 Nov. 1984; and (Part II), 9 Nov. 1984.
216. Quoted in O. Svarlien, 'The Sector Principle in Law and Practice', 10 *Polar Record* 248–63, at 251 (1960); emphasis added.
217. *Ibid.*, emphasis added.
218. See Whiteman, *Digest of International Law*, Vol. 2, at 1262 (1963).
219. *Ibid.*, at 251; emphasis added. See also F. M. Auburn, *The Ross Dependency*, 91 pages (1972), at 52.
220. Art. 6 of Antarctic Treaty.
221. W. M. Bush, *Antarctica and International Law*, Vol. II, at 142–3 (1982); emphasis added.
222. See the opinion of the legal adviser before an Australian Parliamentary Sub-Committee, 12 Oct. 1977, in *ibid.*, at 143, note 4.

223. For the diplomatic correspondence between 1933 and 1938, see Bush, *supra* note 221, at 498–505.
224. *Ibid.*, at 505–6.
225. Reproduced in O. Svarlien, 'The Sector Principle in Law and Practice', 10 *Polar Record* 248–63, at 252 (1960); emphasis added.
226. *Ibid.*
227. Hackworth, *Digest of International Law*, Vol. I, at 463 (1940).
228. *Supra*, note 98, at 311; emphasis added.
229. *Ibid.*, at 611.
230. *Ibid.*, emphasis added.
231. *Ibid.*, at 621.
232. Note of July 1, 1959, M.S. Dep't State file 702.022/6–859, reproduced in Whiteman, *Digest of International Law*, Vol. 2, at 1253.
233. *Ibid.*, at 1244.
234. *Ibid.*, at 1262.
235. *Ibid.*, at 1250. The United States could well claim large areas of the Antarctic on the basis of exploration and other State activities going back some 150 years. It sent an official government expedition as far back as 1838–42, under the command of Lieutenant Charles Wilkes (to wit: Wilkes Land within the Australian and French sectors). In more recent times, Admiral Byrd headed three official expeditions in 1928–30, 1933–5 and 1939–41; see Marie Byrd Land within the unclaimed sector between longitudes 90° and 150° west.
236. *Ibid.*, at 1255.
237. *Ibid.*, at 1254–5; emphasis added.
238. See A. P. Morchan, 'The Legal Status of Antarctica: An International Problem', *Soviet Yearbook of International Law* 356–9, at 357 (1959).
239. For an extensive description of the Soviet activities in the Antarctic, see A. F. Treshnikov, *Découverte et Exploration de l'Antarctique*; Moscow: Editions du Progrès, 529 pages (1972) and B. A. Boazek, 'The Soviet Union and the Antarctic Regime', 78 *A.J.I.L.* 834–58 (1984).
240. *Supra* note 231, at 1261.
241. This presumption is also borne out by Bernier's account of the ceremony in which he states that he referred to 'the granting to Canada by the Imperial Government on September 1st, 1880, of all British *territory in the northern waters* . . .' See J. E. Bernier, *Master Mariner and Arctic Explorer*, 1939, at 128; emphasis added.
242. See Chart 7000, *Arctic Archipelago*, edition of 24 Sept. 1982.
243. 1898 *Stat. Can.*, Chap. 6.
244. On the question of probative value of maps, see in particular the following: *Island of Palmas Case* (1928), United Nations, *Rep. Int'l Arbitral Awards*, Vol. II, 830–71, at 852–4; C. C. Hyde, 'Maps as Evidence in International Boundary Disputes', 23 *A.J.I.L.* 311–16 (1933); *Temple of Preah Vihear Case* (1962) I.C.J. Rep. 3, at 27–9; G. Weissberg, 'Maps as Evidence in International Boundary Disputes: A Reappraisal', 57 *A.J.I.L.* 781–803 (1963); *Indo-Pakistan Western Boundary Case* (1968), United Nations, *Rep. Int'l Arbitral Awards*, Vol. XVII, at 83–99 and 539–43; *Beagle Channel Case* (1977), 52 *Int'l Law Rep.* 93–285, at 201–20 and 229–30 (1979); and B. Bolleaker–Stern, 'L'arbitrage dans l'Affaire du Canal de Beagle', 83 *R.G.D.I.P.* 7–52 (1979).

Part 2
The waters of the Canadian Arctic Archipelago as historic waters

In December 1973, the Legal Bureau of the Department of External Affairs expressed the position of the Canadian government in the following terms: 'Canada . . . claims that the waters of the Canadian Arctic Archipelago are internal waters of Canada, on an *historical basis*, although they have not been declared as such in any treaty or by any legislation'.[1] This is taken to mean that Canada claims to have an historical title to those waters; the waters would be what are commonly called 'historic waters'. Given Canada's position, this part examines the doctrine of historic waters and its applicability to the waters of the Canadian Arctic Archipelago.

5

The basic characteristics of historic waters

This chapter considers four aspects of the doctrine of historic waters: (1) the origin and recognition of historic waters; (2) the legal status of historic waters; (3) the present role of historic waters; and (4) historic waters and related doctrines.

5.1.   Origin and recognition of historic waters

The doctrine of historic waters developed from that of historic bays which had emerged during the 19th century[2] for the protection of certain large bays closely linked to the surrounding land area and traditionally considered by claiming States as part of their national territory. Those bays were often expressed to be of vital importance from the economic and national security standpoints. As rules relating to the delimitation of maritime areas developed, the idea of claiming bays on the basis of an historic title was extended to other areas of the sea adjacent to the coast.

The nature of historic waters in international law was never spelled out in any convention. In 1958, it was the view of the International Law Commission's Special Rapporteur on the Law of the Sea, Mr François, that the Commission did not have sufficient material at its disposal to formulate principles on the matter. Safeguard provisions, however, were inserted in the Convention on the Territorial Sea, thus recognizing the legitimacy of both historic bays and historic waters. Provisions for both cases, however, were formulated as exceptions to the general rules for drawing territorial waters.[3] It was specified that the rules relating to the maximum 24-mile closing line for bays '. . . do no apply to so-called "historic bays" '.[4] In the same way, the equidistance rule governing the delimitation of the territorial sea between States with opposite or adjacent coasts was said not to apply '. . . where it is necessary by reason of historic title . . . to delimit the territorial seas of the two States in a way which is at variance therewith'.[5]

After the 1958 Conference, the Secretariat of the United Nations prepared a document on the 'Juridical Regime of Historic Waters, including Historic Bays',[6] which, along with the 1957 Secretariat document on historic bays, constitutes a good summary of the status of those two doctrines. However, the matter was not pursued by the International Law Commission and the Third Law of the Sea Conference did not address those questions. Consequently, the Convention on the Law of the Sea of 1982 merely reproduced the 1958 provisions without clarifying the legal status of historic waters.[7]

5.2.   Legal status of historic waters

It is now generally accepted that historic waters have the status of internal waters. Even in 1923, when Sir Cecil Hurst made his study on 'The Territoriality of Bays', he concluded that '. . . all the waters lying inwards from this baseline are national waters and form part of the national territory'. He specified that '. . . they stand in all respects on precisely the same footings as the national territory'.[8]

In 1951, the same conclusion was reached by the International Court in the *Fisheries Case* with respect to historic waters generally. The United Kingdom had argued that, although Norway was entitled on historic grounds to claim certain fjords and sounds as internal waters, it was only entitled to claim as territorial waters certain other fjords and sounds which had the character of legal straits.[9] Addressing itself to this argument, the Court began by formulating the principle that 'by "historic waters" are usually meant waters which are treated as *internal waters* but which would not have that character were it not for the existence of an historic title'.[10] At the end of its judgment, it concluded that all of the waters enclosed by straight baselines, including those where an historic title had been established, had the status of internal waters. The Court held specifically that '. . . the waters of the Vestfjord, as indeed the waters of all other Norwegian fjords, can only be regarded as *internal waters*'.[11]

As for the 1958 Convention on the Territorial Sea, it dealt with the legal status of historic bays and historic waters in an indirect way only. It provided that the waters enclosed by the 24-mile closing line for bays 'shall be considered as internal waters'[12] and that the 24-mile rule did not apply to historic bays.[13] The implication is clear that the waters of historic bays are also to be considered as internal waters. And, since no special provision was inserted on historic waters generally, the same legal status must have been intended. In the words of Professor O'Connell, 'It is . . . clear that Article 7(6) should not be interpreted as restricting the category of historic waters to that of historic bays. The latter is but an instance of the former, although, for obvious reasons, the number of claims to bays is greater than to other maritime areas.'[14]

The common provision in the 1958 and 1982 Conventions, touching on historic waters, stipulates that the equidistance rule for the delimitation of the territorial sea between opposite and adjacent coasts is not to apply 'where it is necessary by reason of historic title to use another mode of delimitation'.[15] Since the article deals only with the outer limit of the territorial sea between coasts for which it adopts the equidistance rule, an exception to that rule because of an historic title would simply extend that limit. Consequently, the Conventions would seem to envisage the possibility of an historic title to water areas, aside from bays, resulting only in a status of territorial waters rather than internal waters.

That possibility was also envisaged by the United Nations Secretariat in its study of 1962 when it concluded that the legal status of historic waters 'would in principle depend on whether the sovereignty exercised in the particular case over the area of the claiming State and forming a basis for the claim, was sovereignty as over internal waters or sovereignty as over the territorial sea'.[16] In other words, the exact legal status of historic waters would depend on whether the right of innocent passage has been recognized in those waters. If the claimant State did not allow such passage, the result would be internal waters whereas, if it did, the result would be territorial waters.

As was seen, such an interpretation is possible under both the 1958 and 1982 Conventions. However, it is strongly suggested that, in customary law as interpreted and applied by the International Court in the *Fisheries Case* of 1951, an historic title results in internal waters, whether or not these water areas have the configuration of bays. The status of internal waters means that the sovereignty of the coastal State is complete. Those waters have exactly the same status as 'inland waters' and generally refer to lakes, canals, rivers, ports and harbours. As emphasized by Professor Gidel, the coastal State is no longer obliged to recognize the innocent passage of foreign ships in internal waters.[17] The coastal State may, if it wishes, permit such innocent passage, but is under no legal obligation to do so. If it does, the foreign ship is then exercising a privilege granted by the coastal State, rather than a right recognized by the international community.

Even if such historic waters are enclosed by straight baselines, the right of innocent passage would not exist under the Territorial Sea Convention since those waters would not have been considered part of the territorial sea or of the high seas before their enclosure. In other words, the status of internal waters would not result from the drawing of straight baselines, but rather from the proof of an historic title. This was the opinion expressed by Judge Hackworth in the *Fisheries Case* in the following declaration:

> Judge Hackworth declares that he concurs in the operative part of the judgment but desires to emphasize that he does so for the reason that he considers that the Norwegian Government has proved the existence of an *historical title* to the disputed areas of water.[18]

It would appear that Judge Hackworth did not believe that the disputed areas could be considered as internal waters because of the drawing of straight baselines, but rather because an historic title had been established.

5.3.   Present role of historic waters

The original role of the doctrine of historic title as the sole basis for claiming certain large bays and other wide areas of adjacent waters has been considerably restricted, although preserved, by the 1958 Convention on the Territorial Sea. There are three reasons for this restriction. The first is that the permissible closing line of 24 miles of open water (not counting islands at the entrance) for bays has eliminated the necessity of claiming a number of bays on the basis of an historical title. Secondly, given the appropriate geography, it is now possible to establish straight baselines across indentations, regardless of their configuration and even though their entrances might greatly exceed 24 miles.[19] Thirdly, States may now justify certain baselines, which do not fully meet the geographical requirements, by showing important economic interests of long usage. History is then invoked, not as an autonomous source of title as in the case of historic waters, but as an additional and subsidiary source to reinforce or consolidate the title arising out of the establishment of the straight baseline system. This is how Norway was allowed to rely on an historical consolidation of title in justifying its application of the straight baseline system.

The result of the above is that the doctrine of historic waters may be invoked only when the other three standard bases have proved inadequate to meet the claim of internal waters by the coastal State. In 1982, Judge Oda expressed the limited role of historic waters as follows:

> . . . the concept of historic waters may be claimed only where strict adherence to the geographical conditions required for

internal waters (such as bays, straight baselines) might lead to a somewhat inequitable result because of the longstanding exercise of powers by the coastal State concerned.[20]

In that event, the doctrine of historic waters becomes one of exception to the ordinary rules and, as shall be seen in the next chapter, the burden of proof is more stringent.

5.4. Historic waters and other related doctrines

The doctrine of historic waters has certain common features with the concepts of custom, acquisitive prescription, effective occupation and historical consolidation of title. All of them envisage State activities over a long period of time and require a general toleration on the part of other States. However, the scope of application of those four concepts varies from one to another. *Custom* requires a general practice which is accepted as law by the generality of States and, if it is only a regional custom, a specific acceptance is required on the part of the State against which the practice is invoked. Custom is a source of law in the wide sense of that term and cannot, *per se*, become a source of title. *Acquisitive prescription* properly so-called presumes an adverse holding and normally applies only to territory; it has the effect of transferring to the claimant State a title formerly belonging to another State.[21] *Effective occupation* also is a doctrine which usually applies only to territory, although exceptionally the configuration and position of a bay may be such that it is deemed part of the territory occupied.[22] In a general way, effective occupation might also be said to form the basis of an historic title to both land and water areas in that manifestations of sovereignty for a long period, acquiesced in by foreign States, are required in both cases.

The doctrine of *consolidation of title* is perhaps the most closely related to that of historic waters, in that both concepts apply indifferently to land and water areas. The basic difference is that in the case of historic title to sea areas, history or long usage is invoked as the sole basis of the claim, whereas for a consolidation of title, history serves as only a secondary ground in addition to a primary and standard one such as the straight baseline system. To borrow the words of Professor O'Connell, 'long usage is a ground additional to the other grounds on which a claim may be based – as a confirmation but not as the essence of the claim'.[23] The nature and requirements of an historical consolidation of title will be examined in Chapter 9 on the legal requirements for straight baselines.

Now, to conclude on the specificity of historic waters properly so-called, it is a doctrine which constitutes an exception to the standard rules for the acquisition of title to the maritime domain and, consequently, carries with it stringent requirements.

5.5.   Summary
1. The doctrine of historic waters emerged during the nineteenth century as an enlargement of the doctrine of historic bays and has been preserved in the 1958 and 1982 conventions.
2. Although the conventions seem to envisage the possibility of an historic title resulting in territorial waters, customary law as applied in the *Fisheries Case* of 1951 makes it clear that historic waters are 'internal waters'.
3. The role of historic waters in international law has been considerably reduced, because of the adoption of a 24-mile closing line rule for bays and the introduction of the straight baseline system.
4. The doctrine of historic waters has common features with the concepts of custom, acquisitive prescription, effective occupation and historical consolidation of title, in that they all envisage manifestations of sovereignty for a long period of time by the claimant State and a certain acquiescence on the part of other States.

6

Requirements of historic waters[24]

The conventions are completely silent as to the legal requirements for the existence of historic waters. However, a number of
authoritative studies have been made[25] and it is generally agreed that
three requirements must be met before a claim to historic waters is duly
established. These are: (1) the exclusive exercise of State authority; (2)
long usage or the passage of time; and (3) the acquiescence of foreign
States. In addition to these requirements, three related questions must
also be considered: (1) the legal effect of protest; (2) the vital interests of
the claimant State; and (3) the burden of proof. This chapter will address
those six points in succession.

6.1. Exclusive exercise of State authority

Since a claim to historic waters is one over a maritime area which
the coastal State considers as an integral part of its national territory, the
type of jurisdiction exercised over that area should be essentially the same
as that being exercised on the rest of the territory. More precisely, the
coastal State must exercise an effective control over the maritime area
being claimed to the exclusion of all other States. Naturally, the extent of
control will vary depending on a number of factors such as the size of the
maritime area, its remoteness and the degree of its usability. The actual
control might be limited, but yet sufficient, in remote areas. In the words
of Professor O'Connell, 'just as in the case of remote and little used seas,
very little in the way of effective exercise of sovereignty need be
required',[26] to show that a State took whatever action was necessary to
assert and maintain its authority and control over the area in question.

As to the type of manifestations of sovereignty which must be adduced,
Professor Gidel states that '[t]he exclusion from these areas of foreign
vessels or their subjection to rules imposed by the coastal State, which

exceed the usual scope of regulations made in the interests of navigation, would obviously be acts affording convincing evidence of the State's intent'.[27] The learned author is careful to point out, however, that those are not the only acts of authority which constitute evidence of the exercise of sovereignty by the coastal State. Normally, the physical act of excluding foreign vessels is preceded by national legislation which forbids the entry of foreign ships and subjects them to certain conditons. As an example, Professor Bourquin states, in respect of a bay, that 'the State which forbids foreign ships to penetrate the bay or to fish therein indisputably demonstrates by such action its desire to act as the sovereign'.[28] The same author emphasizes,however, that the intent or desire of the State must be expressed by deeds and not merely by proclamations.[29] This being admitted, if the laws and regulations of the coastal State are never challenged, very little action is necessary to maintain the effective and exclusive control needed to support a claim of sovereignty.

6.2.   Long usage and passage of time

'Historic titles must enjoy respect and be preserved as they have always been by long usage', said the International Court in 1982.[30] How much passage of time must take place before a sufficiently long usage is attained cannot be determined in the abstract. Somewhat as for the formation of a custom, it is impossible to determine in advance how long the effective control must last before it materializes into an historic title. A great variety of terms is employed to describe the length of time required for a usage to have legal effect. The more common expressions used are 'well-established usage', 'continuous usage of long standing', 'continued and well-established usage', 'immemorial usage' and usage 'from time immemorial'.

The length of time required will depend on a number of factors such as the degree of change being effected, the attitude of other States and the political strength of the claimant State. With reference to historic bays, Professor Bourquin states that 'l'usage, dont l'Etat se prévaut en pareil cas, remonte au plus lointain passé. C'est un usage immémorial, au sens propre du mot'.[31] Whether such a long usage is required for historic waters generally is not certain. What is certain is that the longer the effective control endures, the firmer the legal title becomes, resulting in a virtual presumption of general acquiescence.

6.3.   Acquiescence by foreign States

Everybody agrees that the attitude of foreign States is important, particularly on the part of States primarily affected by the usage in question, and that some form of acquiescence is necessary before an historic title can arise. However, there is some disagreement as to the precise form which the acquiescence should take. The opinions seem to be divided into two groups, depending on the view which is taken of the nature of historic waters. Those who consider historic waters as an exception to the general rules relating to the acquisition of maritime sovereignty take a stricter view of acquiescence. They seem to consider acquiescence as a form of consent or recognition of the sovereignty of the coastal State over certain maritime areas, and this recognition or consent must come from those States that are affected by the claim in question. The other group maintains that silence or the absence of protest on the part of the other States is sufficient for the exercise of sovereignty by the claimant State to result in an historic title. The first group, represented by Fitzmaurice, admits that acquiescence need not take the form of a positive act on the part of the foreign States and that the role of the theory of historic rights is to create 'a presumption of acquiescence arising from the facts of the case and from the inaction and toleration of States'.[32] The other group, represented by Professor Bourquin, maintains that while it is false to say that the acquiescence of these States is required, it is true that if their reactions interfere with the peaceful and continuous exercise of sovereignty no historic title can be formed.[33]

The difference in approach as to the exact form which the reaction of foreign States should take was evident in the *Fisheries Case* of 1951. While admitting that the reaction of foreign States constitutes a very important element in the formation of an historic title, even taken in the special sense of consolidation of title, The Norwegian Government rejected the view of the United Kingdom that '. . . le titre historique aurait pour seul fondement l'acquiescement des autres Etats et se confondrait ainsi, substantiellement, avec l'institution juridique de la reconnaissance'.[34] The Norwegian Government went on to say that it considered the absence of reaction on the part of foreign States as being sufficient to confirm the peaceful and continuous character of the usage.[35]

The Court seemed to accept the Norwegian argument when it stated that 'the general toleration of foreign States with regard to the Norwegian Practice is an unchallenged fact'.[36] Having found as a fact that the Norwegian straight baseline system had met with the general toleration of foreign States, the Court went on to hold that Norway was entitled to enforce her system against the United Kingdom. The Court stated in particular:

> The notoriety of the facts, the general toleration of the international community, Great Britain's position in the North Sea, her

own interest in the question, and her prolonged abstention would in any case warrant Norway's enforcement of her system against the United Kingdom.[37]

It thus appears from the judgment of the Court that a general toleration or absence of protest on the part of foreign States suffices for a consolidation of title to materialize. But would it suffice to create an historic title as an exception to the general rules relating to the acquisition of sovereignty? This is doubtful, because the general toleration involved in the *Fisheries Case* was toward the straight baseline system which was found to be, not in derogation of, but rather in accordance with, the general rules for the delimitation of maritime jurisdiction. Consequently, the general toleration resulted in an historical consolidation of title rather than in the actual creation of an historical title. In the words of the Court, 'it is indeed this system (of straight baselines) itself which would reap the benefit of general toleration, the basis of an historical consolidation which would make it enforceable as against all States'.[38] If history had been the sole basis for the claim that the enclosed waters were subject to Norway's sovereignty, a more positive form of acquiescence would probably have been necessary. Professor O'Connell puts it as follows:

> In the case of historic waters, what has to be established is the virtually total toleration of those nations whose interests are clearly affected, because the situation, having its origins in an illegal act which time and absence of opposition alone can validate, is analogous to the subversion of a neighbouring title on land by adverse occupation.[39]

Presuming that there exists that kind of acquiescence or total toleration, an historic title to sea areas might well arise. But what if, on the contrary, there have been protests?

6.4.   The legal effect of protest

An effective protest on the part of interested States would rebut the presumption of acquiescence which would normally arise out of a long period of total toleration. To have its legal effect, however, the protest must be a real one, and must usually be followed by some more forceful steps by the protesting State in order to prevent the formation of an historic title. The protest must be a real one in the sense that the mere raising of an objection in an indirect way would not suffice. For instance, in the *Fisheries Case*, the Court held that the negotiations between the Norwegian and the United Kingdom Governments, which followed the 1911 incident when a British trawler was seized and condemned for having violated the fishing limit of Norway, did not suffice to constitute an effective protest. These discussions related to the two questions: the 4-mile breadth of territorial sea and Norway's sovereignty over Varangerfjord. In the opinion of the Court, those two questions 'were unconnected with the position of baselines' and therefore could not be considered as a formal protest. The Court went on to specify: 'It would appear that it was only in its memorandum of July 27, 1933, that the United Kingdom made a formal and definite protest on this point'.[40] It would therefore appear that a formal objection to the specific points at issue is necessary before it can constitute a valid protest. Fitzmaurice does not seem to be quite that strict as to the form which a valid protest may take when he states:

> Apart from the ordinary case of a diplomatic protest, or a proposal for reference to adjudication, the same effect could be achieved by a public statement denying the prescribing country's right, by resistance to the *enforcement* of the claim or by counteraction of some kind.[41]

O'Connell, on the other hand, believes that international tribunals have insisted on considerably more than a reservation of rights:

> International tribunals in cases of disputed territorial claims have not accepted that a reservation of rights once made is sufficient to prevent their negation. They have examined protests in the context of the behaviour of both parties, and have decided that a protest, in order to be effective, must be made by all reasonable and lawful means, and the nature of the means and the vigour with which the protest is made will depend upon the gravity of the threat and the nature of the rights violated.[42]

If such is the requirement for territorial claims of sovereignty, *a fortiori* it must be for maritime claims of the same nature.

Presuming that a valid protest has been made, this might suffice for a while to prevent the period of time from starting to run, but it usually must be followed by some more forceful action on the part of the protesting State. If such a State is really concerned about the possibility of an historic right materializing, it ought to take all permissible means at its disposal to prevent the practice or exercise of authority in question from developing into an historic title. As stated by Blum, '[t]hese means will usually comprise the active prosecution of the objection through diplomatic negotiations, the arrangement of some kind of *modus vivendi*, or the seeking of a solution by enquiry, mediation or conciliation'.[43] In addition to these possibilities, a State could of course bring the matter to the forum of the United Nations, particularly if the dispute threatens international peace and security. As for the possibility of seeking a judicial or arbitral settlement, the protesting State might be unable to invoke this mode of settlement if the claimant State has not recognized the jurisdiction of the International Court or has continuously refused in the past to submit disputes to arbitration. Naturally, the degree of effectiveness of a protest will depend on a number of factors such as the interest of the protesting State, its geographical situation, its political strength, and the fact that it is not the sole protestor.

6.5.   Vital interests of the coastal State

In addition to relying on the standard requirements of effective control for a long period of time, in the absence of protest and with the acquiescence of foreign States, coastal States will usually buttress their claim of historic waters by invoking vital interests. This is particularly so for a claim of historic bay, where such factors as geography, vital needs of the local inhabitants and national security are invoked. This was done, for instance by the Soviet Union in 1957 in the case of the Bay of Peter the Great, when it invoked the special geographical configuration of that bay as well as its special economic and defence significance.

Although some writers would prefer to treat the question of vital interests as a basis for a claim outside the historic title,[44] the great majority of authorities, including arbitral and judicial tribunals, deal with the question of vital interests along with the requirements for the establishment of an historic title. For instance, Dr L. M. Drago, in his dissenting opinion in the *North Atlantic Coast Fisheries Arbitration* of 1910, quoted with approval from Westlake who maintained that certain bays belong to the coastal State 'when such country has asserted its sovereignty over them, and particular circumstances such as geographical configuration, immemorial usage and above all, the requirements of self-defence, justify such a pretension'.[45] In addition, the arbitral decision itself did admit the importance of the vital interest factor when it stated that 'conditions of national and territorial integrity, of defence, of commerce and of industry are all vitally concerned with the control of the bays penetrating the national coast line'.[46]

The vital interest factor was also given consideration and weight by the International Court in the *Fisheries Case*, particularly with respect to the geographic and economic aspects. The importance of the peculiar geography of Norway permeates the whole judgment of the Court. Even when discussing the strictly historic aspect of Norway's claim, the Court emphasized that the Norwegian system 'was imposed by the peculiar geography of the Norwegian coast'.[47] The Court also referred to the regional economic interests which had been established by long usage in favour of the coastal fishermen of Norway.

The question of defence interests or national security in the case of bays is developed by Mitchell Strohl, a commander in the U.S. Navy, in his book on the International Law of Bays. With respect to the status of Hudson Bay he comes to the conclusion that since the bay is a potential area for operation of submarines, this is a relevant consideration:

> With its tremendous area and location it constitutes a real problem for anti-submarine defense. For this reason Canada may be expected to see Hudson Bay as being more intimately connected with her national security than in times past. If Hudson Bay remains a part of Canadian internal waters, then Canada can, by international law, forbid the entry of foreign warships including submarines. This does not make an intruding submarine any easier to find, but it does permit Canada to take forceful and positive measures against such craft if their presence is detected.[48]

Similar reasoning could, of course, apply to the waters of the Canadian Arctic Archipelago generally, which may easily be penetrated by submarines in spite of the presence of ice.

To summarize the relevance of vital interests in making a claim of historic waters, it may be stated that they have not acquired the status of one of the constitutive elements forming the basis of such a claim, but it is a factor which may be taken into account in the general appraisal of those elements.[49] In a similar way, the special Court of Arbitration, in the *English Channel Continental Shelf Case* between France and the United Kingdom in 1977, stated that vital interests such as security, defence and navigational interests could be taken into account. Although such interests could not have a decisive influence, they could support and strengthen conclusions already indicated by geographical, political and legal considerations.[50]

6.6. Burden of proof of historic waters

The question of whether there exists a special burden of proof in a claim of historic waters depends on the precise nature of 'historic waters'. If the proper nature of a claim of historic waters is one in derogation of, or as an exception to, the general rules for the acquisition of sovereignty over sea areas, the burden is on the claimant State to prove an historic title to those areas. Gidel states that 'en ce qui concerne le fardeau de la preuve, il pèse sur l'Etat qui prétend attribuer à des espaces maritimes proches de ses côtes le caractère, qu'ils n'auraient pas normalement, d'eaux intérieures'.[51] The learned author is even more explicit in a later development when he maintains that 'l'Etat riverain qui prétend à des eaux historiques réclame pour elles un traitement exceptionnel; il faut que ce traitement exceptionnel soit justifié par des conditions exceptionnelles'.[52] If, on the contrary, such a claim is one for the acquisition of sovereignty within the standard rules such as those applicable to the straight baseline system, then there is no such burden of proof on the claimant State.

In the *Fisheries Case*, both the United Kingdom and Norway agreed that the burden of proof rested with the party claiming an exceptional right, but they disagreed as to whether the Norwegian application of the straight baseline system constituted such an exception. The United Kingdom contended that 'the burden of proof lies upon the State which invokes the historic title'.[53] It added that 'the role of the historic element being to validate what is an exception to general rules and therefore intrinsically invalid, it is natural that the burden of proof should so emphatically be placed upon the coastal State . . .'.[54] Norway did not agree that she had such a heavy burden of proof since, in her opinion, the method of delimitation in question did not constitute an exception to the general rules.

> En résumé, le titre historique, tel que le Gouvernement norvégien le conçoit et l'invoque dans le présent litige, n'a aucunement pour rôle de légitimer une situation par ailleurs illégale, mais bien de confirmer la validité de cette situation.[55]

The Court seems to have accepted Norway's argument that she was invoking history not to validate an otherwise illegal situation but rather to confirm the validity of that situation, since it concluded that the straight baseline system used by Norway was not contrary to international law. Professor Charles De Visscher, who participated in the Court's judgment, explained that 'the historic title invoked by Norway was not at all equivalent to a claim over historic waters'.[56] This authoritative interpretation of the Court's judgment is fully in accord with the Court's basic reason for its conclusion, namely that the general toleration of the straight baseline system had resulted in 'an historical consolidation which would make it enforceable as against all States'.[57] This means that the Norwegian claim of sovereignty over the waters landward of baselines was established not on the basis of an historical title to those waters, but rather on an historic consolidation of the straight baseline system as the basis for such sovereignty. Professor O'Connell expresses a similar interpretation of the Court's judgment as follows:

> The Court did not uphold the Norwegian method of drawing the territorial sea upon historic grounds, since this would have been to treat that method as a special case rather than as the application to a concrete situation of standard rules, which is what the Court intended.[58]

In the light of the above, it may be concluded that no special burden of proof rests on a State basing its claim on a consolidation of title but there is such a burden when the basis of the claim is an historic title.

6.7. Summary

1. The basic requirements of an historic title to sea areas are the following:

   (1) exclusive authority and control over the maritime area claimed, including the expulsion of foreign ships if necessary;

   (2) long usage or the passage of a long period of time, the length of the period depending on the circumstances; and

   (3) acquiescence by foreign States, particularly those clearly affected by the claim.

2. An effective protest on the part of interested States will negate acquiescence and prevent an historic title from materializing.

3. The vital interests of the claimant State may be taken into account, but only in making a general appraisal of the basic requirements.

4. When an historic title is the sole basis for a claim of maritime sovereignty, it is a claim in derogation of the general rules for the acquisition of such sovereignty and a special burden of proof rests on the claimant State.

7

State practice of arctic States on historic waters

The doctrine of historic waters not having been spelled out in any convention, its precise legal context remains difficult to circumscribe in customary law and, consequently, State practice becomes even more significant than otherwise would be the case. This is particularly so with respect to States situated in the Arctic, since the claim of historic waters being appraised is to a maritime area of that region. This chapter will examine the State practice of Norway, Denmark, the United States, the Soviet Union and Canada.

7.1. State practice of Norway
In 1951, Norway relied on history, in addition to the straight baseline system, as a basis for its claim that the waters of the fjords and sunds enclosed by the baselines were internal waters. It was invoking history as a subsidiary ground only, not in support of a claim of historic waters but rather one of consolidation of title over some of the areas of the sea enclosed by straight baselines.[59] This consolidation of title was shown to exist in particular with respect to the Lopphavet Basin where the local inhabitants had enjoyed historic fishing and whale hunting rights. Norway does not appear to rely now, any more than it did in the past, on historic waters properly so-called and there is no provision for such waters in its legislation.[60]

7.2. State practice of Denmark
There is no evidence that Denmark has ever made any claim of historic waters. It has used the straight baseline system along the coasts of Greenland and the Faroes, but the waters enclosed are not claimed as historic waters. Denmark has no provision for historic waters in its legislation.

7.3. State practice of the United States

The United States claims an historic title to Delaware Bay and Chesapeake Bay, both of which are only about ten miles wide and which may now be considered as closed under the 24-mile rule of the 1958 Territorial Sea Convention irrespective of any historic title. In 1957, the United States made a formal protest to the Soviet Union when the latter made its claim to the Bay of Peter the Great which has an entrance of 108 miles. On instructions from his government, the American ambassador in Moscow delivered a Note protesting the claim and specifying that 'under international law, the body of water enclosed . . . cannot, either geographically or historically, be regarded as part of the internal waters of the U.S.S.R.'[61] Whether the protest was followed by any further action is not known. In addition the United States has sent Coast Guard icebreakers in all of the seas along the north coast of the Soviet Union between 1947 and 1967.[62] This indicates that the United States does not recognize those seas as internal waters of the Soviet Union, should these or some of them be claimed as historic waters. There is no provision for historic waters in the legislation of the United States.

7.4. State practice of the Soviet Union

Although Soviet writers are virtually unanimous in claiming Siberian seas as historic waters, the Soviet Union itself has never made any such claim and its State practice remains rather unclear.[63] It has, however, made a claim of historic bay and has expressly provided for historic waters in its legislation.

As already mentioned, the Soviet Union claimed the Bay of Peter the Great as an historic bay in 1957 and expressly provided for the exclusion of navigation in the Bay by foreign ships and flights over the Bay by foreign planes except by permission. The Soviet government claimed that 'the waters of the Bay of Peter the Great are *historically* waters of the Soviet Union by force of the special geographic conditions of that bay and its special economic and defence significance'.[64] This claim was protested by a number of countries, particularly by the United States, the United Kingdom, France and Japan.[65] The latter sent three protest Notes but to no avail, and it would appear that the Soviet Union has been able to enforce its claim.

As for its legislation, the Soviet Union adopted a new law in 1960 prior to its ratification of the 1958 Convention on the Territorial Sea, entitled 'Statute on the Protection of the State Border of the U.S.S.R.'[66] The Statute provides that 'internal sea waters of the U.S.S.R. shall include . . . waters of bays, inlets, coves, and estuaries, *seas and straits, histori-*

cally belonging to the U.S.S.R.'[67] In 1982, a new law on the State
Boundary of the U.S.S.R. was passed and the provision just quoted was
retained.[68] In 1984, the Council of Ministers adopted a decree declaring
Penzhinskaya Inlet historic waters, closing it with a 38.2 mile line and, in
1985, claimed as 'waters historically belonging to the U.S.S.R.' the
following sea and bays: the White Sea, Cheshskaya Bay and
Baydaratskaya Bay. The closing lines measure 84.4, 44 and 62.5 nautical
miles respectively.[69] The above legislation could certainly envisage the
Siberian seas (Kara, Laptev, East Siberian and Chukchi Seas) and the
straits joining these seas (Vilkitski, Sannikov, Dimitri Laptev and De
Long Straits). However, it would not appear from Soviet State practice
that the seas and straits mentioned are considered as historic waters.[70]
Certain events which took place in 1965 and 1967, involving American
Coast Guard icebreakers, would seem to indicate that Kara Sea at least is
not regarded as historic internal waters. In September 1965, the *Northwind* did come under surveillance by Soviet aircraft and a warship when it
entered the Kara Sea, but it nevertheless spent nearly two months in that
sea, taking oceanographic readings and collecting water samples throughout most of the sea, including the southern part between Novaya Zemlya
and the coast. The Soviet destroyer kept the American ship under
surveillance, but it never actually interfered. The Soviet government is
reported to have objected to the bottom-coring programme carried out
by the *Northwind* as being contrary to the 1958 Continental Shelf
Convention;[71] however, that objection could well have been justified
since the Convention provides for the express consent of the coastal State
for exploration activities.[72]

   As for the Vilkitski Straits, leading out of the Kara Sea and into the
Laptev Sea, the Soviet Union did strongly protest the intended American
crossing in 1965 but apparently not on the ground that they were historic
straits. A spokesman from the State Department is reported to have
stated that the Soviet government had emphasized 'that they consider the
waters of Vil'kitskogo Strait *territorial waters*, and that anyone intruding
in those waters would be intruding on the territory of the Soviet Union.
They made it very clear they would go all the way to take appropriate
measures'.[73] Although the waters were claimed to be territorial waters
only and the right of innocent passage should normally have applied, the
Soviet Union was technically in a position to refuse passage since its
reservation to the Territorial Sea Convention did not recognize this right
to warships and provided for prior authorization.[74] It would appear that
the Soviet Union was correct in classifying the American icebreakers as
warships since they probably came within the definition of warships in the

High Seas Convention. However, its refusal to allow passage was questionable in both customary and conventional law. The second incident involves the American icebreakers *Eastwind* and *Edisto*, which explored the same Kara Sea in August and September 1967 after being refused permission to go through the Vilkitsky Straits. In the words of Captain Bankert of the *Eastwind*, 'we put in a profitable seven weeks in the Kara Sea and made a very complete oceanographic survey of the entire sea, which had only been partially surveyed before by United States vessels'.[75] The refusal to allow passage through the Vilkitsky Straits was based, as in 1965, on the ground that 'the straits constituted Soviet *territorial* waters'.[76] According to Captain Benkert, 'the Russians replied that they felt, since the strait was their territorial waters and since we were warships, a 30-day notice to pass through these waters was requested'.[77]

It seems rather clear from the above that the Soviet Union does not claim Kara Sea nor the Vilkitsky Strait as historic internal waters. It would also appear that she adopts a similar view with respect to Chukchi Sea, East Siberian Sea and Laptev Sea, since all of the seas were surveyed by surface ships of the United States, particularly in the years 1947, 1962, 1963 and 1964.[78] In addition, the United States conducted under-ice experiments in the Chukchi Sea at least three times: in 1947, with the *Boarfish* (SS-327); in 1948, with the *Carp* (SS-338);[79] and in 1960, with the U.S.S. *Sargo* (SSN-583).[80]

In spite of the above, Soviet jurists are virtually unanimous in claiming Siberian seas as internal waters, mainly on the basis of history. A 1961 international law textbook, edited by Professor Kozhevnikov, summarizes the opinion of Soviet writers as follows:

> Soviet jurists rightly consider the Siberian seas which are akin to gulfs (Kara Sea, Laptev Sea, East Siberian Sea and Chukotsk Sea) as Soviet internal waters. These seas, which in effect constitute gulfs and which are of exceptional economic and strategic importance to the Soviet Union have over a *prolonged period of history* been used by Russian seafarers.[81]

It will be noted that history or long usage is invoked as a basis, along with geography (the seas being considered as gulfs), economics and national security.

In 1966, the Soviet naval international law manual claimed two of the three straits linking the Siberian seas as historic straits:

> The Dimitri Laptev and Sannikov Straits are regarded as *belonging to the Soviet Union historically*. They have never been used for international navigation, and in view of specific natural

conditions and frequent ice jams, the legal status of these straits is sharply distinguished from all other straits being used for international navigation.[82]

Unlike the Vilkitsky Straits which are 11 and 22½ miles wide, causing an overlap of territorial waters, the Dimitri Laptev and Sannikov Straits both have a width of about 29 miles.

In 1976, three Soviet jurists (F. S. Boitsov, G. G. Ivanov and A. L. Makovski) published a book, 'The Law of the Sea', in which they claim all four Siberian seas as historic waters:

> The legal status of the waters of the Arctic Ocean is very special. Kara Sea, Laptev Sea, East Siberian Sea and Chukchi Sea (within the Soviet sector) are situated in the Arctic Ocean. These seas are the seas of the bay type and *belong historically to the U.S.S.R.*[83]

In 1978, Professor V. F. Meshera, a well-known authority in maritime law at the University of Leningrad, published a book entitled 'The Legal Status of Maritime Areas' in which he claims that three of the Siberian seas are historic waters:

> The bays and seas situated near the coast of a single State may be included in the territory of that State irrespective of the width of their entrance, if it is justifiable by virtue of the *vital interests* of the coastal State and there exists an established practice of so doing. These are called historic seas. Kara Sea, Laptev Sea and East Siberian Sea are such *historic seas.*[84]

In additon to history, the learned writer invokes the vital interests of the coastal State as the Soviet Union had done to buttress its claim to the Bay of Peter the Great in 1957.

To sum up on the State practice of the Soviet Union it may be stated that, although there exists legislation which provides for historic seas and straits, the government does not appear to have made any such claim in the Arctic. Soviet jurists, on the other hand, are unanimous in claiming at least three of the Siberian seas (Kara Sea, Laptev Sea and East Siberian Sea) as historic seas. In addition, two straits (Dimitri Laptev and Sannikov) are claimed to belong to the Soviet Union historically.

7.5.   State practice of Canada

It is now established Canadian government policy that the waters of the Canadian Arctic Archipelago are claimed as historic internal waters. This is not clear from Canadian legislation, but it has become clear in official statements made since the voyage of S/T *Manhattan* across the Northwest Passage in 1969.

Canadian legislation does not make any express reference to historic waters. However, the definition of internal waters incorporated in the Territorial Sea and Fishing Zones Act of 1964 is wide enough to include such waters. The Act provides that 'the internal waters of Canada *include* any areas of the sea that are on the landward side of the baselines of the territorial sea of Canada'.[85] Since the definition is inclusive and not exclusive, it could well include maritime areas which are internal waters, not because they are on the landward side of baselines,[86] but on the basis of geography or history, or both. Bodies of water which are or may be so considered include a number of bays off the coast of Newfoundland, Nova Scotia, Prince Edward Island, New Brunswick and Quebec,[87] the Bay of Fundy,[88] the Gulf of St Lawrence,[89] Dixon Entrance and Hecate Strait,[90] Queen Charlotte Sound,[91] Hudson Bay and Hudson Strait[92] as well as the waters of the Canadian Arctic Archipelago.

As for government policy, the first official statement which indicated that Canada might be claiming the waters of the Canadian Arctic Archipelago as historic internal waters was that of the Prime Minister at the time of the *Manhattan* crossing in October 1969. The statement was part of a prepared address on the Speech from the Throne and read in part as follows:

> Canadian activities in the northern reaches of this continent have been far-flung but pronounced for many years, to the exclusion of the activities of any other government. The Royal Canadian Mounted Police patrols and administers justice in these regions on land and ice, in the air and *in the waters*.[93]

Having specified that the Canadian Eskimos pursue 'their activities over the icy waters without heed as to whether that ice is *supported* by land or *by water*', the statement emphasizes the long duration of those activities and concludes by saying that 'Arctic North America has, for 450 years, progressively become the Canadian Arctic'.[94]

In December 1969, after the voyage of the *Manhattan*, the Standing Committee on Indian Affairs and Northern Development took a clear position on the status of the Arctic waters in its Report to the House of Commons. The Report stated: 'Your Committee considers that the waters lying between the islands of the Arctic Archipelago have been, and are, subject to Canadian sovereignty *historically*, geographically and geologically.'[95]

In December 1973, an official of the Department of External Affairs replied to a letter as to the legal status of Arctic waters and specified that Canada claimed, as historic internal waters, not only Hudson Bay and Hudson Strait but also the waters of the Canadian Arctic Archipelago.

On Hudson Bay and Strait he quoted the Minister of Northern Affairs who had replied to a question in the House of Commons in 1957, saying that 'the waters of Hudson Bay are Canadian waters by historic title . . .' and that Canada regarded 'as inland waters all the waters west of a line drawn across the entrance to Hudson Strait from Button Island to Hatton Head on Resolution Island'.[96] The Department's letter continued: 'Canada also claims that the waters of the Canadian Arctic Archipelago are internal waters of Canada, *on an historical basis*, although they have not been declared as such in any treaty or by any legislation.'[97] This unquestionably constitutes the clearest and most precise statement as to the nature of and basis for Canada's claim over Arctic waters.

It is obvious from the above that Canada considers the waters of the Canadian Arctic Archipelago as historic internal waters, at least since 1973 and perhaps since 1969 or before.

7.6.   Summary

1. In 1951 Norway invoked history, but not the doctrine of historic waters, as a subsidiary basis, along with the straight baseline system, for its claim that the waters of the fjords and sounds along its coast were internal waters. Norway has no provision in its legislation for historic waters.

2. Denmark has made no claim of historic waters and there is no provision in its legislation for such waters.

3. The United States has claimed a few bays (both about 10 miles wide) as historic bays, but has protested strongly a claim of historic bay (108 miles wide) by the Soviet Union. It has no provision for historic waters in its legislation.

4. The Soviet Union claims the Bay of Peter the Great as an historic bay since 1957 and expressly provides for historic seas and straits in its legislation. However, it does not appear to have claimed the Siberian seas and straits along its coasts as historic waters. Soviet writers, on the other hand, have been virtually unanimous in claiming these seas as historic internal waters. In addition, the Soviet naval international law manual claims the Dimitri Laptev and Sannikov Straits as belonging to the Soviet Union historically.

5. Canada has claimed a number of bays as historic bays and, since 1973 at least, claims all the waters of the Canadian Arctic Archipelago as historic internal waters. It has no express provision for historic waters in its legislation, but its definition of internal waters is wide enough to include historic waters.

8

Historic waters applied to the Canadian Arctic Archipelago

The object of this chapter is twofold: first, to examine State activities of Great Britain and Canada in relation to the waters of the Canadian Arctic Archipelago and second, to appraise Canada's claim of historic waters in light of the legal requirements for the establishment of such a claim.

8.1. State activities of Great Britain before the transfer in 1880

The instructions to the official expeditions of the British explorers was to discover a passage between the Atlantic and Pacific Oceans.[98] An examination of the journals and narratives of those expeditions also reveals that their takings of possession in what is now the Canadian Arctic Archipelago were confined to land.[99] This conclusion simply confirms an earlier finding by Dr Gordon Smith in his historical study of those explorations that 'they were concerned with land, not water; and it is conspicuous that whatever claims were made these claims were for the most part specifically to land'.[100] Nevertheless, it becomes pertinent to review briefly the activities of Great Britain in the waters of the Archipelago, as Canada's predecessor in title, with a view to determining the extent of its presence in those waters before the transfer of the Arctic islands to Canada in 1880. If the review were to conclude that Great Britain's presence did not extend to much of the waters and that they were often frequented by foreign ships, it would become somewhat more difficult for Canada to meet the requirements of exclusive control for the necessary long period of time and the general acquiescence of foreign States.

British explorers, beginning with Martin Frobisher in 1576 and ending with those in search of the Franklin expedition in 1859, covered virtually all the waters of the Canadian Arctic Archipelago. A glance at a couple of maps, one on the period from 1576 to 1847 (see Figure 12) and the other from 1847 to 1859 (see Figure 13) is sufficient to gain a reasonably accurate idea of the waters explored. The explorations, at times carried out on the ice and on land along the coasts, were particularly intense during the search for the Franklin expedition between 1847 and 1859. During that period alone, over seventy search expeditions were dispatched to the Canadian Arctic, about half of which spent at least one winter and some of them up to three. Although none of the British explorers actually completed the navigation of the Northwest Passage, they effectively discovered all of its various routes.

As to the degree of exclusiveness of the British presence, all but one of the expeditions in the waters of the Canadian Archipelago were British. The exception was the American expedition headed by De Haven in 1850, which reached Wellington Channel in search of Sir John Franklin's expedition. Along with British explorers, British whalers began to follow the whale into Lancaster Sound and sometimes further westward beginning around 1800. Although Dutch, German and American whalers were quite active in the Davis Strait/Baffin Bay area during the eighteenth and the beginning of the nineteenth centuries, it seems that only the British whalers ventured in the waters of the Archipelago itself.[101]

8.2.   State activities of Canada after the transfer in 1880

In 1880, Great Britain transferred to Canada all her British territories and possessions in North America not already included within the Dominion of Canada, as well as all the islands adjacent to any such territories or possessions with the exception of the Colony of Newfoundland. Shortly after the transfer, Canada began sending expeditions to the Canadian Arctic with a view to consolidating its title to the islands and exercising control over the water areas, particularly in relation to whaling vessels. In 1897, the Wakeham expedition was sent to the Frobisher Bay and Cumberland Sound areas with a view to asserting sovereignty where a few American whalers had been. The instructions to William Wakeham were 'to proceed up the Sound (Cumberland), to take as formal possession of the country as possible, to plant the flag there as notice that the country is ours, and take all necessary precautions to inform natives and foreigners that the laws must be observed and particularly the customs laws of Canada'.[102] Pursuant to his instruction, Dr Wakeham did make a formal declaration of possession at Kekerten in Cumberland Sound, in the presence of the Scottish whaling agent and his own crew and 'the flag was hoisted as an evidence that Baffin's Land with all the territories, islands and dependencies adjacent to it were now, as they always had

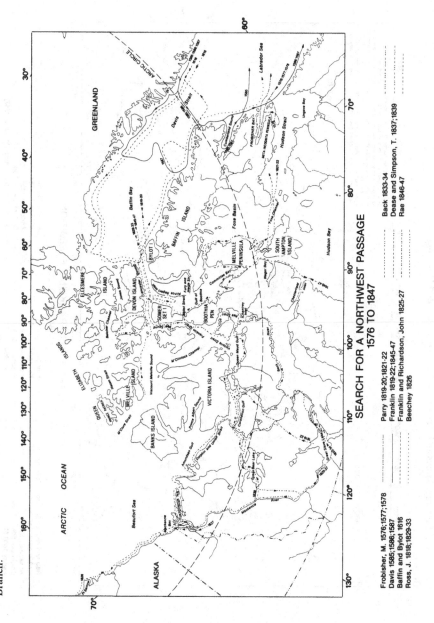

SEARCH FOR A NORTHWEST PASSAGE
1576 TO 1847

Frobisher, M. 1576;1577;1578
Davis 1585;1586;1587
Baffin and Bylot 1616
Ross, J. 1818;1829-33

Parry 1819-20;1821-22
Franklin 1819-22;1845-47
Franklin and Richardson, John 1825-27
Beechey 1826

Back 1833-34
Dease and Simpson, T. 1837;1839
Rae 1846-47

115

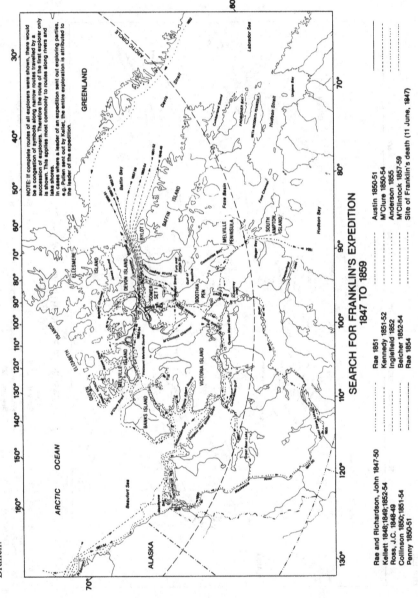

Fig. 13. Search for Franklin's expedition, 1847 to 1859. Reproduced with permission of Energy, Mines and Resources, Canada Surveys and Mapping Branch.

GREENLAND

ARCTIC OCEAN

ALASKA

SEARCH FOR FRANKLIN'S EXPEDITION
1847 TO 1859

NOTE: If complete routes of all explorers were shown, there would be a congestion of symbols along narrow routes travelled by a succession of explorers. Therefore the route of the first explorer only is shown. This applies most commonly to routes along rivers and lake shores.
In cases where a leader of an expedition sent out exploring parties, e.g. Pullen sent out by Kellet, the entire exploration is attributed to the leader of the expedition.

Rae and Richardson, John 1847-50
Kellett 1848;1849;1852-54
Ross, J.C. 1848-49
Collinson 1850;1851-54
Penny 1850-51

Rae 1851
Kennedy 1851-52
Inglefield 1852
Belcher 1852-54
Rae 1854

Austin 1850-51
M'Clure 1850-54
Anderson 1855
M'Clintock 1857-59
Site of Franklin's death (11 June, 1847)

116

been since their first discovery and occupation, under the exclusive sovereignty of Great Britain'.[103]

In 1903, Canada sent an expedition headed by A. P. Low 'for the purpose of patrolling, exploring and establishing the authority of the Government of Canada in the waters and islands of Hudson Bay, and the north thereof'.[104] Low was accompanied by S. W. Bartlett, Master of the *Neptune*, Major J. D. Moodie of the Northwest Mounted Police and five other members of the force, as well as a crew of twenty-nine and a scientific staff of five. Low was given the powers of a justice of the peace to enforce fisheries laws and Major Moodie was appointed customs officer and magistrate. The expedition spent its first year in the Hudson Bay region and, in the summer of 1904, proceeded to the Northwest Passage as far as Port Leopold on the south side of Barrow Strait where Low hoisted the Canadian flag and left a copy of the Canadian Customs regulations in the *Investigator*'s boiler abandoned by Sir James Clark Ross in 1849.[105]

The Low expedition was followed by a series of four expeditions in quick succession from 1904 to 1910, all of them with the C.G.S. *Arctic* under the command of Captain J. E. Bernier. The 1904–5 expedition led by J. D. Moodie, patrolled Hudson Bay and established a divisional police headquarters at Churchill.[106] The 1906–7 expedition, headed by Captain J. E. Bernier, sailed on July 28, 1906, shortly after the adoption of an amendment to the Fisheries Act, with instructions to collect whaling dues. The amendment provided for a licence fee of $50.00 payable by any whaling vessel within the waters of Hudson Bay or the territorial waters north of the 55th parallel. The amendment was as follows:

> Notwithstanding anything in this section, the license fee payable for any vessel or boat engaged in the whale fishery or hunting whales within the waters of Hudson Bay, or the *territorial waters of Canada* north of the 55th parallel of north latitude, if not so engaged or hunting in connection with a factory established in Canada, shall be fifty dollars for each year; and, insomuch as Hudson Bay is wholly *territorial water of Canada*, the requirements of this section . . . shall apply to every vessel . . . in any part of the waters of Hudson Bay, whether such vessel or boat belongs to Canada, or is registered and outfitted in . . . any other British or foreign country.'[107]

It will be noted that the expression 'territorial water(s) of Canada' is used to qualify the waters of Hudson Bay and those north of the 55th parallel. It must be remembered that, although the concept of territorial waters was well established in international law at that time, it was not so for the concept of 'internal waters'. Even in 1923 Sir Cecil Hurst referred to the waters of an historic bay as 'national waters', explaining that 'they stand in all respects on precisely the same footings as the national territory'.[108] Whether the intention was to treat the waters north of the 55th parallel in the same way as those of Hudson Bay is not altogether clear from the language of the amendment. The wording of the whaling licence, however, would appear to make a distinction, for it states that the vessel 'is hereby licensed to engage in the Whale Fishery or Hunting Whales within the waters of Hudson Bay, or the *territorial waters of Canada* north of the 55th parallel of north latitude'.[109] Did this mean that a licence was necessary only if a vessel wanted to hunt whales within Hudson Bay and the three-mile territorial waters of Canada measured from the mainland and around each island? If so, foreign vessels could hunt whales without a licence in all of Lancaster Sound as well as in Barrow Strait except within three miles of the islands lying in the middle of the western portion of that strait. Be that as it may, Captain Bernier does not appear to have limited himself to the three miles in his enforcement of the licence requirement.

In his account of this voyage Bernier states that he left notices and copies of the new law on whaling licences with an agent for distribution at Albert Harbour, Pond Inlet, and then travelled to Lancaster Sound and Barrow Strait where he personally issued licences to whalers. He did so particularly from Whaler Point, at Port Leopold, which he described as a well-known place of call for all ships sailing in those waters.[110] During this expedition, Bernier made several formal takings of possession of islands on both sides of Parry Channel. He also took possession of Lowther Island in the western portion of Barrow Strait.

On July 28, 1908, Captain Bernier led a second expedition to the Arctic waters and proceeded as far west as Mercy Bay in M'Clure Strait. According to Bernier himself, he was sent to the Arctic for the purpose 'of patrolling the waters contiguous to that part of the Dominion of Canada already annexed, and for the purpose of annexing territory of British possessions as far west as longitude 141°'.[111] He states in his report that 'specific instructions were given as to the *waters to be patrolled*, explored, and lands to be annexed in continuation of the two voyages already made to the northern waters by the same ship, commanded by myself'.[112] It was during this voyage, on July 1, 1909, that Captain Bernier made a formal taking of possession of the whole Arctic Archipelago, between longitude 60° and 141° west, up to the North Pole, by placing a memorial plaque on Parry's Rock at Winter Harbour, Melville Island. Also, during this voyage, licences were sold to Scottish whalers and at least one American, Harry Whitney of New York. Captain Bernier describes his encounter with Mr Whitney on September 5, 1909 as follows:

> I informed Mr Whitney that I was patrolling Canadian waters, and, as he had on board his vessel a motor whale boat, it would be necessary for him to take out a fishery licence, and that I would issue it. He stated that if it was a regulation, he would pay the legal fee of $50, and take the licence. I accordingly issued the licence and received the fee.[113]

In the expedition of 1910–11, again led by Captain Bernier, he received the following instructions from the Deputy Minister of Marine and Fisheries:

> You will acquaint any persons who you may find engaged in the whale fishery in these northern waters that you are *patrolling these waters* as the duly accredited officer of the Canadian Government, and you will, where necessary, demand payment of licence fees for such fishing. If payment be refused, you will make a request that such refusal be put in writing. It is not desirable that you should take any action in this regard which would be likely to embarrass the Government.[114]

It would appear from the above that it was the feeling of the Government that Captain Bernier had been a little forceful in the past in implementing Canadian whale hunting regulations. Nevertheless, the important fact is that he received specific instructions to patrol northern waters and enforce whale fishery regulations in those waters. The instructions specified that the waters to be patrolled included Davis Strait, Baffin Bay, Lancaster Sound, Barrow Strait, Viscount Melville Sound, M'Clure Strait and the Beaufort Sea as far as Herschel Island.[115]

Captain Bernier sailed to Pond Inlet, patrolled the various inlets north and west of Baffin Island, then proceeded west in Parry Channel as far as M'Clure Strait. Unable to proceed further west because of the ice, he came back to Pond Inlet. During this expedition, Captain Bernier carried out his duties as fishery officer and issued whaling licences. In particular, he boarded the Newfoundland ship *Diana*, in August 1910, accompanied by his customs officer Lavoie.[116]

In 1913, the Canadian Government decided to send an official government expedition, under the command of Vilhjalmur Stefansson. The purposes of the expedition are set out in considerable detail in two Orders in Council. The first one, dated February 22, 1913, specified that 'the expedition will conduct its *explorations in waters and on lands* under Canadian jurisdiction or included in the northern zone contiguous to the

Canadian territory'.[117] The Order further specified that 'the expedition would also have occasion to examine into the operations of the American whalers which frequent the northern waters of Canada, and of putting into force the Customs and Fisheries Regulations which these whalers should observe'.[118] The Order in Council finally stated that the scientific studies would include soundings of the continental shelf. The second Order in Council, dated June 2, 1913, issued immediately prior to the departure of Stefansson's expedition from Victoria, stated that the purpose of the expedition was 'for a scientific exploration of northern lands, and of the Arctic seas to the north of the Delta of the MacKenzie'.[119] In this Order in Council, Stefansson was given 'authority to take possession of and annex to His Majesty's Dominions any lands lying to the north of Canadian territory which are not within the jurisdiction of any civilized power'.[120]

During this expedition, which lasted from 1913 to 1918, Stefansson discovered a number of new islands in the northwestern part of the Archipelago and of which he took possession in the name of Canada. He also explored the greater portion of the Beaufort Sea, including the taking of numerous soundings, thus outlining the continental shelf from Alaska to Prince Patrick Island.[121]

After World War I Canada felt that further expeditons should be made to the Arctic on a regular basis in order to affirm and maintain its sovereignty and an annual patrol was institutionalized. This began in 1922 and became known as the Eastern Arctic Patrol. For the first four years the patrol was sent out with the C.G.S. *Arctic*, always under the command of Captain J. E. Bernier. The patrols were carried out every year at least until 1958 and extended throughout the Canadian Arctic. The R.C.M.P. were involved in the majority of those patrols and they established posts in numerous places, particularly on the outskirts of the Archipelago. The annual patrol and the permanent presence of R.C.M.P. posts in the more important areas thus consolidated Canada's sovereignty over the islands and insured a reasonable degree of supervision over the adjacent waters.[122]

In addition to the Arctic expeditions and patrols just described very briefly, Canada exercised control over certain supply missions to Arctic waters by American ships after the war. Although Canada granted permission to the United States to establish a number of weather stations on some of its Arctic islands in 1948, it did not have sufficient icebreakers at that time to insure the annual sea supply missions to those stations. Consequently, it was agreed that the U.S. Navy should request approval in advance for the routes to be followed. In a similar way, in 1957 during the construction of the D.E.W. (Distant Early Warning) line installations by the United States along the seventieth parallel, a squadron of American icebreakers was led to its destination, across the narrow Bellot Strait, by the Canadian naval icebreaker H.M.C.S. *Labrador*.

At that time Prime Minister St Laurent was careful to explain to the House of Commons the precautionary measures which had been taken by Canada, when permitting U.S. vessels to carry on hydrographic surveys in order to ensure the sea supply to the D.E.W. line stations. Not only were there always Canadian representatives aboard, he said, but 'each year the United States Navy has been required to apply for a waiver of the provisions of the Canada Shipping Act, since the cargo ships they charter operate in Canadian coastal waters'.[123] On the same occasion and in answer to a question by Mr Green as to whether the government considered those waters as 'Canadian territorial waters', Mr St Laurent answered: 'Oh, yes, the Canadian government considers that these are *Canadian territorial waters*, and we make it a condition of the consent we have given to these arrangements that they apply for a waiver from the provisions that would otherwise apply in Canadian territorial waters.'[124] It is very obvious from the context that the term 'territorial waters' was being used as synonymous with the modern expression 'internal waters'.

Aside from the three icebreakers which actually completed a crossing of the Northwest Passage in 1957, eight more American ships completed a transit of the Passage. However, as will be seen in Part 4, it seems that all of them crossed the passage with Canada's express or implied consent. A limited number of foreign ships, however, are presently using part of the Northwest Passage to transport zinc concentrates from Nanisivik Mine south of Lancaster Sound and Polaris Mine north of Barrow Strait. Those ships do not appear to be subjected to any requirement of prior authorization and presumably are exercising their right of innocent passage. This would be in conformity with various assurances given by Canada at the time of the *Manhattan* voyage in the Northwest Passage in 1969 and the enactment of the Arctic Waters Pollution Prevention Act in 1970. Canada made clear at that time that its concern was the protection of the marine environment and that, if this was sufficiently guaranteed, it would not interfere with the passage of foreign ships.[125]

8.3. Appraisal of Canada's claim of historic waters

As was seen in Chapter 6, a claim to historic waters is one in derogation of the general rules for the acquisition of maritime sovereignty, and the requirements are stringent. Canada would have to prove that she has exercised exclusive authority over the waters of the Archipelago for a long period of time and that such authority has been acquiesced in by foreign States, particularly by those clearly affected by the claim. Such claim has to be judged in the light of any formal protest that might have been made and the practice of States, especially those in the region.

On the positive side, it may be stated that, prior to the transfer of the islands to Canada in 1880, virtually all of the waters of the Canadian Arctic Archipelago were discovered by British explorers and frequented practically only by them and British whalers. After the transfer, Canada patrolled most of those waters, beginning with the more southern ones such as Hudson Bay and Strait, Frobisher Bay and Cumberland Sound, and then extending its patrols to Lancaster Sound, Barrow Strait and the connecting inlets and sounds to the south. It adopted legislation in 1906, requiring whalers to obtain a licence when hunting whales in Hudson Bay and the territorial waters north of the 50th parallel. This legislation was enforced until the end of whaling in Arctic waters, around 1915. Indeed, whaling licences appear to have been issued for whaling beyond the limits of territorial waters.

In 1922, the Eastern Arctic Patrol was instituted and annual patrols were made until at least 1958. These patrols extended occasionally to the western Arctic waters and were carried out mostly by the R.C.M.P. In 1926, the Arctic Islands Preserve was adopted to protect the natives and wildlife, and to indicate that Canada controlled the area within the sector formed by the 60th and 141st degrees of longitude. This was followed in 1929 by the Game Regulations applicable in the Preserve.

After World War II, the Canadian Coast Guard was established. Its main functions in the Arctic consist of icebreaking services and the resupply of Arctic communities. It has provided icebreaking services in particular for the few foreign transits of the Northwest Passage which have taken place so far, including that of the *Manhattan* in 1969. The Coast Guard is also charged with the implementation of the regulations relating to pollution prevention and shipping safety control adopted under the Arctic Waters Pollution Prevention Act of 1970.

Since 1970, Canadian survey ships have been active in surveying and charting the waters of the Archipelago, particularly the straits which are expected to be used for the transportation of hydrocarbons from the Beaufort Sea and the Arctic islands. In 1977, Canada instituted the NORDREG reporting system which provides for all ships to report to the Coast Guard before entering the waters of the Archipelago.

On the negative side of Canada's claim of historic waters, it must be realized that both British and Canadian explorers confined their takings of possession to lands and islands. Even the formal taking of possession by Captain J. E. Bernier on July 1st, 1909 'of the whole Arctic Archipelago lying to the north of America from longitude 60° W to 141° W up to latitude 90° N'[126] has to be interpreted as being limited to the land areas. Bernier himself stated in his report covering that expedition that 'specific instructions were given as to the waters to be patrolled, explored, and *lands to be annexed*'.[127] In addition, the sector theory which is implicit in the formulation of this taking of possession is of no legal value as a basis for a claim of sovereignty in international law, even if such claim is restricted to lands and islands.[128]

Also on the negative is the considerable doubt that exists as to whether the whale hunting legislation of 1906, making the purchase of a licence mandatory, applied to all waters north of the 55th parallel aside from Hudson Bay. The wording of the licence itself would seem to limit the requirement to the 'territorial waters of Canada',[129] which extended to three miles at that time.

A further and more conflicting aspect of Canada's claim of historic internal waters is found in some of the official statements made in 1970 when Canada's territorial waters were extended from three to twelve miles. Explaining the implication of this extension when moving a second reading of the Bill, the Secretary of State for External Affairs explained that 'the effect of this bill on the Northwest Passage is that under any sensible view of the law, Barrow Strait, as well as the Prince of Wales Strait, are subject to complete Canadian sovereignty'.[130] This could perhaps indicate that the Minister thought this amendment would make the waters internal, although it is difficult to imagine how. Indeed, in answering a question on straight baselines the day before, he had stated that 'since obviously we claim these to be Canadian internal waters we would not draw such lines'.[131] Be that as it may, the intended effect of the amendment must have been to create an overlap of territorial sea in the western portion of Barrow Strait where lies a string of five islands in a zig-zag fashion across the strait. The widest passage being only 15.5 miles, between Lowther and Young Islands, there would now be a sort of gate of territorial waters across Barrow Strait, as there was already in Prince of Wales Strait where the Princess Royal Islands, lying in mid-strait, reduce the width of the passage to less than six miles. This intended effect of the amendment was made abundantly clear a few days later by the legal adviser of the Department of External Affairs, when testifying before the Standing Committee on External Affairs and National Defence. He stated, in particular:

This has implications for Barrow Strait, for example, where the 12-mile territorial sea has the effect of giving Canada sovereignty from shore to shore. To put it simply, we have undisputed control – undisputed in the legal sense – over two of the *gateways* to the Northwest Passage.[132]

In other words, even if a foreign ship succeeded in avoiding Prince of Wales, as the *Manhattan* had attempted to do in 1969 by entering M'Clure Strait instead, it could no longer remain on a strip of high seas or of exclusive economic zone. In effect, the extension of territorial waters to 12 miles, partly with a view to creating an overlap of such waters in Barrow Strait, may constitute an admission that the rest of the waters of Parry Channel were considered as high seas.[133] And, of related significance, is the fact that it was not until three years later, in 1973, that the Legal Bureau of External Affairs claimed that the waters of the Archipelago were '*internal waters*' of Canada, on an historical basis'.[134]

In addition to the above, always on the negative side, it must be recalled that the United States made a formal protest in 1970 not only against Canada's extension of its territorial sea to 12 miles, but also against the Arctic Waters Pollution Prevention Act adopted at the same time. This second piece of legislation enabled Canada to enforce certain pollution prevention standards of construction, manning and equipment against all ships navigating in the waters of the Archipelago north of the 60th parallel and up to a distance of 100 miles outside the Archipelago. The United States' protest Note stated that 'international law provides no basis for these proposed unilateral extensions of jurisdictions on the high seas, and the U.S.A. can neither accept nor acquiesce in the assertion of such jurisdiction.[135] The Note ends by suggesting to Canada that the matter be submitted to the International Court of Justice for adjudication. Canada ignored the protest Note as it related to the extension of the territortial sea, but not so with respect to pollution prevention. On that occasion, the Prime Minister stated categorically: 'In short, where we have extended our sovereignty, we are prepared to go to court. On the other hand, where we are only attempting to control pollution, we will not go to court until such time as the law catches up with technology.'[136] Indeed, on the same day the Government introduced the Bill on Arctic waters pollution prevention, the Canadian Ambassador to the United Nations transmitted a letter to the Secretary-General, modifying Canada's acceptance of the International Court's jurisdiction and excepting from such jurisdiction 'disputes arising out of or concerning jurisdiction or rights claimed or exercised by Canada . . . in respect of the prevention or control of pollution or contamination of the marine environment in marine areas adjacent to the coast of Canada'.[137]

The damaging part of this reservation in relation to Canada's claim of historic waters is that the 'marine areas adjacent to the coast of Canada', as described in the new legislation, cover not only a strip of 100 miles outside of the Archipelago but also all of the waters within the Archipelago north of the 70th parallel. If these waters had really been considered as internal waters of Canada, over which it claimed as complete a sovereignty as it did over the lands and islands of the Archipelago, there would have been no doubt as to Canada's jurisdiction to adopt such legislation for the waters within the Archipelago. Thus, the reservation could have been limited to the strip of 100 miles outside the Archipelago and along the northern coast of the Yukon and the Mackenzie Delta. As it was, the reservation indicated an uncertainty on the part of Canada as to the legal basis of this legislation, not only as it applied to the waters outside but apparently also as to the waters inside the Archipelago. Consequently, Canada cannot be said to have ignored the protest of the United States but, on the contrary, seems to have acted upon it.

Another negative element in relation to Canada's claim of historic waters is the fact that the NORDREG reporting system is a voluntary one only. And, indeed, if there is any doubt that the waters of the Archipelago are internal waters, it would be difficult for Canada to insist that foreign ships abide by the reporting system or be refused entry into the Northwest Passage. Presuming that a foreign ship conforms with the Arctic Waters Pollution Prevention Act of Canada, the validity of which should now be considered as confirmed by customary international law,[138] it should have a right of innocent passage in the waters of the Northwest Passage. If these waters are not internal, they are at best territorial waters and innocent passage applies.

For all of the above reasons, the conclusion is that Canada would not succeed in establishing that the waters of the Canadian Arctic Archipelago are historic internal waters.

Notes to Part 2

1. 12 *C.Y.I.L.*, at 279 (1974); emphasis added.
2. See Memorandum prepared by the U.N. Secretariat on Historic Bays, A/CONF. 13/1 (1957).
3. In the words of the International Court, the references to historic bays and historic titles in those provisions are made 'in a way amounting to a *reservation* to the rules set forth therein'; see *Tunisia/Libya Continental Shelf Case* (1982) I.C.J. Rep., at 74, emphasis added.
4. Convention on the Territorial Sea and the Contiguous Zone, A/CONF. 13/L.52 (1958), Art. 7, para. 6.
5. *Ibid.*, Art. 12, para. 6.
6. See A/CONF. 4/143 (March 1962).
7. See A/CONF. 62/121: Art. 10, para. 6 on historic bays and Art. 15 on historic waters.
8. Sir Cecil Hurst, 'The Territoriality of Bays', 3 *B.Y.I.L.* 42, at 54 (1923).
9. [1951] I.C.J. Rep., at 130.
10. *Ibid.*, emphasis added.
11. *Ibid.*, at 142, emphasis added.
12. 1958 Convention, Art. 7, para. 4; and 1982 Convention, Art. 10, para. 4.
13. 1958 Convention, Art. 7, para. 6; and 1982 Convention, Art. 10, para. 6.
14. D. P. O'Connell, *The International Law of the Sea,* Vol. I, at 420 (1982).
15. 1958 Convention, Art. 12, para. 1; and 1982 Convention, Art. 15.
16. 'Juridical Regime of Historic Waters, including Historic Bays', A/CONF.4/143, at 23 (March 1962). The International Court addressed this point briefly in the *Tunisia/Libya Continental Shelf Case* and stated that 'general international law . . . does not provide for a *single* "regime" for "historic waters" or "historic bays", but only for a particular regime for each of the concrete, recognized cases of "historic waters" or "historic bays".' See (1982) I.C.J. Rep., at 74. It is submitted that this *obiter dictum* is of doubtful validity and the more accurate view of the doctrine of historic waters resulting in internal waters was expressed by Judge Oda in that same case, at 196.
17. Gidel, *Le Droit international public de la mer,* Vol. III, at 625 (1934).
18. [1951] I.C.J. Rep., at 144; emphasis added.
19. *Supra* note 4, Art. 4.
20. *Continental Shelf Case, Tunisia/Libya* (1982) I.C.J. Rep., at 210.
21. See in particular the following: W. E. Beckett, 'Les Questions d'intérêt général . . .', *Recueil des Cours,* Vol. 50, at 248 (1934); D. H. N. Johnson, 'Acquisitive

Prescription in International Law', 27 *B.Y.I.L.* 332–54 (1950); Charles De Visscher, *Theory and Reality in International Law*, at 200 (1957); and O'Connell, *supra* note 14, at 426–7.

22. See *The Direct United States Cable Co. Ltd. v. The Anglo–American Telegraph Co. Ltd. et al.* (1877) 2. A.C. 394, at 419.
23. *Supra* note 14, at 422.
24. These requirements have been examined before by this writer in *The Law of the Sea of the Arctic* (1973) at 106–117, and this chapter draws upon that previous examination.
25. See in particular the following: 'Juridical Regime of Historic Waters, including Historic bays', A/CONF.4/143 (March 1962), which reviews State practice, arbitral and judicial decisions, codification projects and the opinion of writers; and O'Connell, *supra* note 14, at 427–38.
26. *Supra* note 14, at 428.
27. *Supra* note 17, at 633, as translated and reproduced in 'Juridical Regime of Historic Waters, including Historic Bays', A/CONF.4/143 (March 1962) and 1962 *I.L.C. Yearbook*, Vol. II, at 14.
28. *Supra* note 6, at 14.
29. M. Bourquin, 'Les baies historiques', in *Mélanges Georges Sauser-Hall*, at 49 (1952).
30. [1982] I.C.J. Rep., at 73.
31. *Supra* note 29, at 49.
32. Fitzmaurice, 'The Law and Procedure of the International Court of Justice 1951–1954; General Principles and Sources of Law', 30 *B.Y.I.L.* 1, at 30 (1953).
33. See Bourquin, *supra* note 29, at 46.
34. [1951] I.C.J. Pleadings, Vol. III, at 462.
35. See *ibid.*
36. [1951] I.C.J. Rep., at 138.
37. *Ibid.*, at 139.
38. *Supra* note 36; words in brackets added.
39. *Supra* note 14, at 39.
40. *Supra* note 36.
41. *Supra* note 32, at 42, footnote 1.
42. *Supra* note 14, at 41. In support of his statement, O'Connell refers to the following: *Chamizal Arbitration*, 5 *A.J.I.L.* 782 (1911); *Minquiers and Ecrehos Case* (1953) I.C.J. Rep., at 47; and *Fisheries Case* (1951) I.C.J. Rep., at 116.
43. Y. Z. Blum, *Historic Titles in International Law*, at 162 (1965).
44. See Bourquin, *supra* note 29, at 51, where he states: 'Le titre historique est une chose; l'intérêt vital en est une autre'.
45. L. M. Drago, *North Atlantic Coast Fisheries Case*, Scott's Hague Court Reports 195, at 200 (1961).
46. *Ibid.*, 141, at 139.
47. [1951] I.C.J. Rep., at 139.
48. M. P. Strohl, *The International Law of Bays*, at 250 (1963).
49. For a similar conclusion, see O'Connell, *supra* note 14, at 437–8.
50. Lauterpacht, *International Law Reports*, Vol. 54, at 98, para. 188 (1979).
51. Gidel, *supra* note 17, at 632.
52. *Ibid.*, at 651. For a similar conclusion see Y. Z. Blum, *supra* note 43, at 232.
53. (1951) I.C.J. Pleadings, Vol. II, at 645.
54. *Ibid.*,
55. (1951) I.C.J. Pleadings, Vol. III, at 461.
56. Charles De Visscher, *Problèmes d'interprétation judiciaire en Droit international public,* at 176, footnote 2 (1963); my translation.

57. [1951] I.C.J. Rep., at 138.
58. O'Connell, *supra* note 14, at 206.
59. This is discussed fully in Chapter 9.
60. Such legislation is not necessary but, if it existed, it might indicate a possible reliance on the doctrine.
61. Whiteman, *Digest of International Law*, Vol. 4, at 254 (1965).
62. See *Polar Record*: 195, at 38; 1965, at 732–3; and 1966, at 305–6.
63. On the State practice of the U.S.S.R. generally, see: Wm E. Butler, *The Soviet Union and the Law of the Sea* (1971); S. M. Olenicof, *Territorial Waters in the Arctic: the Soviet Position*, 52 pages (July 1972); and Wm E. Butler, *Northeast Arctic Passage* (1978).
64. *Supra* note 61, at 255; emphasis added. The same provision was retained in the new Statute 'Law on the State Boundary of the U.S.S.R.', entered into force on March 1, 1983; see English translation in 22 *Int'l Legal Materials* 1055–76, at 1057 (1983).
65. *Ibid.*, at 251–4.
66. *Soviet Statutes and Decisions*, Vol. III, No. 4, at 9 (Summer 1967) and S. H. Lay *et al.* (eds), *New Directions in the Law of the Sea*, Vol. I, at 30 (1973).
67. *Ibid.*, Art. 4, para. c; emphasis added. The same provision was retained in the new Statute 'Law on the State Boundary of the U.S.S.R.', entered into force on March 1, 1983; English translation in 22 *Int'l Legal Materials* 1055–76, at 1057 (1983).
68. A possible exception related to Dimitri Laptev and Sannikov Straits which are mentioned in a publication of the State Department as having been described by the Soviet Union in an 'Aide Mémoire of July 21, 1964' as 'internal waters'; see U.S. State Dept, *Limits on the Seas*, No. 36, 3rd ed. rev., at 37 (1977).
69. See Richard Petrow, *Across the Top of Russia*, at 344–5 (1967) for map showing the location and number of water sample stations.
70. *Ibid.*, at 351.
71. See Art. 2, para. 2 of the Continental Shelf Convention, 1958.
72. *Supra* note 68, at 352–3; emphasis added.
73. The authorization procedure is set out in the 'Rules for Visits by Foreign Warships to Territorial Waters and Ports of the U.S.S.R.', adopted on 25 June 1960, and provides for permission to be requested at least 30 days before the proposed visit; see *Soviet Statutes and Decisions*, Vol. III, No. 4, at 45 (Summer 1967).
74. *The Polar Times*, at 13 (Dec. 1967).
75. 57 *Department State Bulletin*, No. 1473, at 362 (1967); emphasis added. For a discussion of the incident, see this writer's Note 'Soviet Union Warns United States Against Use of Northwest Passage', 62 *A.J.I.L.* 927–35 (1968).
76. *Supra* note 74.
77. See 13 *Polar Record* 305–6 (1966).
78. 'Submarines in the Arctic', 87 *U.S. Naval Institute Proceedings*, No. 10, 58–65 (Oct. 1961).
79. A. S. McLaren, 'The Development of Cargo Submarines for Polar use', 21 *Polar Record* 369–81 (1983).
80. F. I. Kozhevnikov (ed.), *International Law*, at 206 (1961); emphasis added.
81. Quoted by Wm E. Butler, *Northeast Arctic Passage*, at 86 (1978); emphasis added. These straits are also considered by P. D. Barabolia as being historic straits; see *id.*, at 87.
82. See *supra* note 68.
83. F. S. Boitsov, G. G. Ivanov and A. K. Makovski, *The Law of the Sea*, at 49–50 (1976). The writer is indebted to Professor Lucia Galenskaya, Faculty of Law of

the University of Leningrad, for the translation of this passage and the one which follows.

84. V. F. Meshera, *The Legal Status of Maritime Areas*, at 6–7 (1978); emphasis added.

85. S. 3(2) *Statutes of Canada*, 1962, c.22; emphasis added.

86. For a good discussion on this point, see Lawrence L. Herman, 'Proof of Offshore Territorial Claims in Canada', 7 *Dal. L.J.* 3–38, at 4–11 (1982).

87. For a list of those bays, see 12 *C.Y.I.L.* at 277–8 (1974).

88. For a study of the status of the Bay of Fundy, see G. V. la Forest, 'Canadian Inland Waters of the Atlantic Provinces and the Bay of Fundy Incident', 1 *C.Y.I.L.* 149 (1963). See also J. Y. Morin, 'Les eaux territoriales du Canada en regard du Droit international', 1 *C.Y.I.L.* 82–148, at 104–6 (1963).

89. See Press Communiqué dated 6 March 1975, reproduced in 14 *C.Y.I.L.* at 324 (1976). For a prior doctrinal discussion of the status of the Gulf, see J. Y. Morin, 'Le progrès technique, la pollution et l'évolution récente du droit de la mer au Canada, particulièrement à l'égard de l'Arctique', 8 *C.Y.I.L.* 158–248, at 173–83 (1970).

90. see J. Y. Morin, *ibid.*, at 186–9. See also C. B. Bourne and D. M. McRae, 'Maritime Jurisdiction in the Dixon Entrance: The Alaska Boundary Re-examined', 14 *C.Y.I.L.* 175–223 (1976).

91. *Ibid.*, at 189–92.

92. See J. Y. Morin, *supra* note 87, at 183–4 and letter of Legal Bureau of the Department of External Affairs in 12 *C.Y.I.L.* at 278–9 (1974). It should be noted that a baseline across Hudson Strait was established in 1937 by Order in Council P.C. 3134.

93. *Can. H. C. Debates*, Vol. I, at 39 (24 Oct. 1969); emphasis added.

94. *Ibid.*, at 40; emphasis added.

95. *Proceedings of Standing Committee on Indian Affairs and Northern Development*, No. 1, at 6 (16 Dec. 1969); emphasis added.

96. 12 *C.Y.I.L.* at 279 (1974).

97. *Ibid.*, emphasis added.

98. See in particular the instructions to the following explorers: Sir John Ross, in *A Voyage of Discovery*. . . (1819); W. E. Parry, in *Parry's Voyage 1819–20* (1821), at XIX; W. E. Parry, in *Parry's Second Voyage 1821–2–3* (1924), at XXI; W. E. Parry, in *Parry's Voyage 1824–5* (1926), at XVII; and Sir John Franklin in 1845, in *Arctic Blue Books*, Vol. I, Document No. 45216, para. 1.

99. See in particular *Parry's Voyage 1819–20* (1821), at 74 and Sir John Ross, *Narrative of a Second Voyage 1829–33* (1835), at 116 and 156. For a summary of British expeditions from 1821 to 1845, see also 'Chronological List of Explorations and Historical Events in Northern Canada 1821–45', 16 *Polar Record*, No. 100, 41–61 (1972).

100. Gordon Smith, *A Historical Survey of Maritime Exploration in the Canadian Arctic and its Relevance with Subsequent and Present Sovereignty Issues*, typescript, 34 pages, at 26 (1970).

101. For an excellent study, see W. Gillies Ross, 'The Annual Catch of Greenland (Bowhead) Whales in Waters North of Canada 1719–1915: A Preliminary Compilation', 32 *Arctic* 91–121 (1979).

102. See statement by the Minister of Marine and Fisheries in *Can. H. C. Debates*, col. 1816 (6 May 1897).

103. W. Wakeham, *Report of the Expedition to Hudson Bay and Cumberland Gulf in the Steamship 'Diana'*, Ottawa, at 71–8 (1898).

104. Instructions to Major Moodie reproduced in A. E. Millward, *Southern Baffin Island*, Report by the Department of the Interior, at 14–15 (1939).

105. Andrew Taylor, *Geographical Discovery and Exploration in the Queen Elizabeth Islands*, 172 pages, at 112 (1955).
106. A. Cooke and C. Holland, *The Exploration of Northern Canada*, at 302 (1978).
107. *Statutes of Canada*, 1906, 5 Edw. VII, c.13; R.S.C. 1906, c.45, s.12; emphasis added.
108. *Supra* note 8.
109. Emphasis added.
110. See J. E. Bernier's, *Master Mariner and Arctic Explorer*, at 312 and 313 (1939).
111. See Bernier's letter dated April 5, 1910, addressed to the Deputy Minister of Marine and Fisheries, accompanying his report, in J. E. Bernier, *Cruise of the Arctic 1908-9*, at xix (1910).
112. *Ibid.*, at 1; emphasis added.
113. *Ibid.*, at 273.
114. Instructions dated 5 July 1910, reproduced in the Preliminary Part of Bernier's Report, published as *The 'Arctic' Expedition, 1910*, 161 pages (1910); emphasis added.
115. *Ibid.*
116. *Ibid.*, at 85.
117. P.C. 406, 22 February 1913.
118. *Ibid.*
119. P.C. 1316, 2 June 1913.
120. *Ibid.*
121. See P.C. 2887, 21 January 1921, which indicates the main accomplishments of the expedition. See also map reproduced in A. Taylor, *supra* note 105, at 121.
122. For a thorough study of the Eastern Arctic patrol and the role of the R.C.M.P., see Gordon W. Smith, *An Historical and Legal Study of Sovereignty in the Canadian North and Related Problems of Maritime Law*, Part A, Volume II, part of an ongoing study still unpublished.
123. *Can. H. C. Debates*, 6 April 1957, at 3185-6.
124. *Ibid.*
125. See, in particular, a statement of Prime Minister Trudeau in *Can. H. C. Debates*, Vol. I, at 39 (1969) and the Canadian Note to the United States reproduced as an appendix to *Can. H. C. Debates*, at 6028 (17 April 1970).
126. See photograph of plaque with inscription in J. E. Bernier, *Master Mariner and Arctic Explorer*, at 128 (1939).
127. *Supra*, note 112; emphasis added.
128. See the conclusion of Part 1.
129. *Supra* note 109.
130. *Can. H. C. Debates*, at 6015 (17 April 1970).
131. *Ibid.*, at 5953 (16 April 1970).
132. Standing Committee on External Affairs and National Defence, *Minutes of Proceedings and Evidence*, No. 25, at 18; emphasis added.
133. Today, the rest of the waters would be considered part of the exclusive economic zone.
134. See 'Canadian Practice' in 12 *C.Y.I.L.*, at 279 (1974); emphasis added.
135. See Text of U.S. Press Release as Appendix 'A' in *Can. H. C. Debates*, at 5923 (15 April 1970).
136. P. E. Trudeau, 'Canada Leads Fight Against Pollution', *Statements and Speeches*, No. 70/3, at 2 (15 April 1970).
137. Text of letter reproduced in 22 *External Affairs*, at 130-1 (1970).
138. See Donat Pharand, 'La contribution du Canada au développement du droit international pour la protection du milieu marin: le cas spécial de l'Arctique', 11 *Etudes internationales* 441-6 (1980).

Part 3
The waters of the Canadian Arctic Archipelago and straight baselines

The straight baseline system of delimiting territorial waters was developed by Norway, beginning with its first Royal Decree of 1812. The system was approved by the International Court in 1951, and incorporated in the Convention on the Territorial Sea of 1958 and in the Law of the Sea Convention of 1982.[1] Under this system, where a coast is deeply indented or is bordered by an Archipelago, it is permissible to draw straight baselines, across the indentations and between the outermost points of the islands, from which the territorial sea is measured. This geographic situation is commonly referred to as that of a coastal Archipelago.

At the Third Law of the Sea Conference, the applicability of the straight baseline system was extended to mid-ocean or oceanic Archipelagos. These are situated at such a distance from any mainland as to be considered an independent whole and constitute the national territory of a State called an Archipelagic State. Off-lying and oceanic Archipelagos forming an integral part of the territory of a continental State are not specifically provided for in the 1982 Convention.

In 1974, a group of nine States introduced a Working Paper proposing the right for a coastal State, with an off-lying Archipelago forming an integral part of its territory, to apply the straight baseline system to such an Archipelago, subject to the right of innocent passage.[2] Also in 1974, Ecuador proposed that straight baselines be applicable 'to Archipelagos that form part of a State, without any change in the natural regime of the waters of such Archipelagos or of their territorial sea'.[3]

The Informal Single Negotiating Text of 1975 did not include any such provision and limited itself to saying that the provisions relating to Archipelagic States were 'without prejudice to the status of oceanic Archipelagos forming an integral part of the territory of a continental State'.[4] Even this saving clause disappeared in the subsequent texts of the Conference. As a consequence, the 1982 Convention covers only two basic categories of Archipelagos: islands bordering a coast and a group of islands in mid-ocean constituting the national territory of a State.

It would appear from the above that, regardless of their precise configuration and nature, Archipelagos must be inserted in one of two categories envisaged and described in the 1982 Convention. Fortunately, the Convention contains the usual saving provision for customary law, incorporated in law-making treaties of general application, that 'matters not regulated by this Convention continue to be governed by the rules and principles of general international law'.[5] Consequently, if the rules for Archipelagos found in the Convention are too narrow to fit precisely a particular Archipelago, resort may be had to customary law. It might prove necessary to do this in appraising the applicability of the straight baseline system to the Canadian Arctic Archipelago.

In any event, it is advisable here to determine as accurately as possible what is the state of customary law on coastal Archipelagos, since the 1982 Convention is not yet in force and the 1958 Convention is in force only between less than 50 States. Canada in particular, whose Arctic Archipelago is being appraised, is not a Party to the 1958 Convention. In these circumstances, it becomes more important than usual to assess State practice. This Part will examine the various questions just raised under the following headings: the law on straight baselines for coastal Archipelagos; State practice; and the application of straight baselines to the Canadian Arctic Archipelago.

9

The law on straight baselines for coastal Archipelagos

9.1. Geographical requirements for a coastal Archipelago

The geography required for the application of the straight baseline system was laid down by the International Court of Justice in the *Fisheries Case* of 1951. Having stated that the breadth of the territorial sea should be measured from the low-water mark, the Court examined three methods of implementing the low-water mark rule: the *tracé parallèle*, the arcs of circles, and the straight baseline system. It was in its discussion of the method of the *tracé parallèle* that the Court, in effect, described the kind of coast required for the application of the straight baseline system.

> Where a coast is deeply indented and cut into, as is that of Eastern Finmark, or where it is bordered by an Archipelago such as the 'skjaergaard' along the western sector of the coast here in question, the base-line becomes independent of the low-water mark, and can only be determined by means of a geometrical construction.[6]

Two observations should be made about this passage. First, the straight baseline system is made applicable to two types of coast:' *where it is deeply indented or where it is bordered by an Archipelago*'. Of course, a coast could have both of those characteristics either in whole or in part. The other observation relates to the second type of coast, that is 'where it is *bordered by an archipelago* such as the "skjaergaard" '. It could appear that this type of coast is somewhat different from a simple 'fringe of islands along the coast in its immediate vicinity', as provided for by the Conventions,[7] so that customary law as formulated by the Court would require only that there be an Archipelago close to its border, and the Norwegian skjaergaard is given as an example of an appropriate kind of Archipelago. The skjaergaard is constituted of some 120,000 insular formations carved out of a mainland coast, broken by large and deeply

indented fjords, obliterating any clear dividing line between the mainland
and the sea. Some of the islands are located at some 60 miles from the
nearest peninsula on the mainland.[8]

On the other hand, the Conventions require that the islands constitute
a fringe in the immediate vicinity of the coast. The ordinary meaning of
'fringe', according to the Oxford dictionary, is 'a border or edging,
especially one that is broken or serrated'. The term is reasonably accurate
to describe the 'skjaergaard' but is somewhat narrower than the geographic situation envisaged. It has been properly pointed out that numerous coastal archipelagos to which the straight baseline system has been
applied 'could only be questionably described as "fringes" '.[9]

Assuming that a group of islands constitutes a fringe, the Conventions
require that they be in the 'immediate vicinity' of the coast. Vicinity being
synonymous to proximity and similar expressions having received an
extensive interpretation by international tribunals,[10] presumably the
expression 'immediate vicinity' would also receive a wide interpretation.
Considering the foregoing and in spite of apparent differences in wording,
it would seem that the Conventions may be interpreted as a simple
codification of the customary law formulated by the Court.

9.2.   Legal nature of straight baselines

Given the appropriate geography for the use of straight baselines,
the question arises whether such baselines should be considered as
exceptions to the general rule of the low-water mark along the sinuosities
of the coast or simply an adaptation of that general rule. Having described
the skjaergaard where baselines become independent of the low water
mark, the International Court provided the following answer:

> In such circumstances the line of the low-water mark can no
> longer be put forward as a rule requiring the coastline to be
> followed in all its sinuosities. *Nor can one characterize as excep-
> tions to the rule the very many derogations which would be
> necessitated by such a rugged coast*: the rule would disappear
> under the exceptions. Such a coast, viewed as a whole, calls for
> the application of a different method; that is, the method of
> baselines . . .[11]

The straight baseline system is thus viewed as a particular application of
the rule of the low-water mark. This is made abundantly clear later in the
judgment when the Court said that it was 'unable to share the view of the
United Kingdom Government, that "Norway, in the matter of baselines,
now claims recognition of an exceptional system" '. It added that 'all that

the Court can see therein is *the application of general international law to a specific case'*.[12]

In spite of the above clear statements to the contrary by the Court, the International Law Commission incorporated provisions in the Territorial Sea Convention of 1958 which made the method of straight baselines an exception to the low-water mark, the latter being called the 'normal baseline'. Indeed, the Convention states that 'except where otherwise provided in these articles, the *normal baseline* for measuring the breadth of the territorial sea is the *low-water line along the coast* . . .'.[13] This article is then followed immediately by the provision on straight baselines already discussed.

Since the same provisions were kept in the 1982 Convention,[14] there does appear to be a difference between the customary law as explained by the Court and the rule incorporated in the Conventions. Or perhaps the exception is more apparent than real, as suggested by Professor O'Connell.

Although Article 4 looks as if it is an exception to the 'normal' rule in Article 3, in fact the normal rule is that of the general direction of the coast, and that rule is exceptionally applied in the circumstances mentioned in Article 4, where the coast is exceptional. Looked at in this way, *the straight baseline method is not something special but a specific instance of a single principle*.[15]

Whether it is possible to extract such an interpretation from Article 4 on the applicability of straight baselines, by reference to the 'general direction of the coast' criterion for their actual application, is not beyond question. However, what is absolutely certain, as explained by Judge De Visscher who participated in the decision, the Court did not view the straight baseline system as an exception but rather 'as an *adaptation of the law made necessary by local conditions'*.[16]

9.3. Mode of application of straight baselines

In order to insure the international validity of straight baselines, their mode of application or actual construction must follow certain criteria. These are intended as guidelines and may be adapted to diverse situations. In the words of the Court, 'certain basic considerations inherent in the nature of the territorial sea, bring to light *certain criteria* which, though not entirely precise, can provide courts with an adequate basis for their decisions, *which can be adapted to the diverse facts in question'*.[17] The Court spelled out three such criteria: (1) the general direction of the coast; (2) the close link between the land and the sea; and

136   *Waters of Arctic Achipelago and straight baselines*

(3) certain economic interests evidenced by long usage. All three criteria were incorporated in the 1958 and 1982 Conventions without change and using the formulations found in the judgment.

General direction of the coast

The judgment of the Court states that 'while . . . a State must be allowed the latitude necessary in order to be able to adapt its delimination to practical needs and local requirements, *the drawing of base-lines must not depart to any appreciable extent from the general direction of the coast*'.[18] The words emphasized were incorporated without any change in Article 4 of the 1958 Convention and Article 7 of the 1982 Convention.[19]

As the Court itself indicated, the criteria in general and this one in particular lend themselves to subjective judgment in their implementation. What constitutes an appreciable departure from the general direction of the coast is a matter on which the coastal State must be allowed a reasonable degree of latitude. In the same way, the Court had previously stated that 'the method of base-lines . . ., within reasonable limits, may depart from the physical line of the coast'.[20] In its application of this first criterion and in refuting the argument of the United Kingdom that the line across the Lopphavet Basin did not respect the general line of the coast because it was situated some 19 miles from the nearest point of land, the Court answered that 'the divergence between the baseline and the land formations is not such that it is a distortion of the general direction of the Norwegian coast'.[21] In the same context, the Court readily admitted that the criterion of general direction is 'devoid of any mathematical precision'.[22]

In arriving at its conclusion on the Lopphavet line the Court formulated the following test:

> In order properly to apply the rule, regard must be had for the relation between the deviation complained of and what, according to the terms of the rule, must be regarded as the *general* direction of the coast. Therefore, one cannot confine oneself to examining one sector of the coast alone, except in a case of manifest abuse; nor can one rely on the impression that may be gathered from a large scale chart of this sector alone.[23]

In other words, the general direction of the coast is determined by examining a small scale map and, except in a case of manifest abuse, looking at the coast as a whole. If such an examination reveals no distortion of the *general* direction, the first criterion is satisfied. It is important to note that the Court itself emphasized the qualifier 'general', to indicate the imprecision of this criterion.

Close link between land and sea
There must be a close relationship between the land and the sea areas which are enclosed. 'The real question raised in the choice of base-lines', said the Court, 'is in effect whether certain *sea areas lying within these lines are sufficiently closely linked to the land domain to be subject to the regime of internal waters*'.[24] The words emphasized were incorporated without change in the Conventions.

The Court specified that this close link was a 'fundamental considera-tion' and the reason is obvious, since the enclosed waters will acquire the status of internal waters over which the coastal State will have as complete a sovereignty as it does over its land areas. In other words, even the right of innocent passage will not apply to the enclosed waters. Nevertheless, the criterion 'should be liberally applied in the case of a coast, the geographical configuration of which is as unusual as that of Norway'.[25] And the Court did apply the criterion rather liberally to the Lopphavet and Vestfjorden areas. The overall ratio of sea to land areas within the Norwegian Archipelago was 3.5 to 1.

As already indicated, the Conventions of 1958 and 1982 reproduced literally this second criterion but they made an important change as to the resulting legal regime of internal waters. Although the enclosed waters are internal in principle, they will be assimilated to territorial waters and subject to the right of innocent passage if they had previously been considered as part of the territorial sea or of the high seas (or exclusive economic zone).[26] Consequently, when the right of innocent passage is preserved, the reason for the intimate relationship between the land and sea areas is considerably lessened and the application of the second requirement should be correspondingly liberalized.

Regional economic interests evidenced by long usage
The two geographical criteria just reviewed are both mandatory in the implementation of the straight baseline system. When straight baselines meet those two criteria, they are validly established. However, in order to add to the probative value of such criteria, it is permissible to take into account 'certain economic interests peculiar to a region, the reality and importance of which are clearly evidenced by a long usage'.[27] Again the Conventions have retained the key words of the judgment and express this criterion as follows:

> Where the method of straight baselines is applicable . . . , account may be taken, in determining particular baselines, of economic interests peculiar to the region concerned, the reality and the importance of which are clearly evidenced by long usage.[28]

In the *Fisheries Case*, the Court invoked this third consideration to reinforce its conclusion, with respect to the 62-mile line (44 plus 18 miles on either side of a submerging rock) across the Lopphavet basin, that 'the divergence between the base-line and the land formations is not such that it is a distortion of the general direction of the Norwegian coast'[29] and, therefore, valid. However, the Court continued, 'even if it were considered that in the sector under review the deviation was too pronounced, it must be pointed out that the Norwegian Government has relied upon an historic title clearly referable to the waters of the Lopphavet, namely, the exclusive privilege to fish and hunt whales granted at the end of the 17th century to Lt-Commander Erich Lorch under a number of licences'.[30]

As recalled by the Court in its judgment, the Norwegian Government was not relying upon history to justify a claim to areas of the sea which the general law would deny but, in the words of its counsel, 'it invokes history, together with other factors, to justify the way in which it applies the general law'.[31] The Court was satisfied that the historical data produced by Norway lent 'weight to the idea of the survival of traditional rights reserved to the inhabitants of the Kingdom over fishing grounds included in 1935 delimitation, particularly in the case of Lopphavet'.[32] The Court concluded by specifying how such rights could be taken into account in validating a particular line.

> Such rights, founded on the vital needs of the population and attested by very ancient and peaceful usage, may legitimately be taken into account in drawing a line which, moreover, appears to have been kept within the bounds of what is moderate and reasonable.[33]

By this use of the historic fishing and hunting rights of the local population to add probative value to the line across the Lopphavet, the Court was also justifying the length of that line.

9.4.   Length of straight baselines

Invoking the analogy of an alleged 10-mile rule for the closing lines of bays, the United Kingdom Government had argued that the length of straight baselines should not exceed ten miles. The Court rejected the argument. Firstly, it held that the alleged 10-mile rule for bays adopted by certain States had 'not acquired the authority of a general rule of international law'.[34] Secondly, and similarly, with respect to a possible 10-mile rule for straight baselines, the Court found that 'the practice of States does not justify the formulation of any general rule of law'.[35] Specifically, the Court held that 'the attempts that have been made to subject groups of islands or coastal archipelagoes to conditions analogous to the limitations concerning bays (distance between the islands not exceeding twice the breadth of the territorial waters, or 10 or 12 sea miles), have not got beyond the stage of proposals'.[36]

The Court did not deem it necessary to say any more on the question of length, except to point out that sometimes several lines could be envisaged and, in such cases, 'the coastal State would seem to be in the best position to appraise the local conditions dictating the selection'.[37] The Court was therefore satisfied that, if a straight baseline can be justified under the two geographic criteria and possibly also under the economic criterion, the line would be valid regardless of its length. In the case of Norwegian Archipelago, the 47 baselines varied from a few hundred yards to what is in effect 62 miles across the Lopphavet, the 62-mile line running 44 and 18 miles on either side of a submerging rock.

Both the International Law Commission and the 1958 Conference on the Law of the Sea were satisfied that, having regard to the criteria, no maximum length for straight baselines should be adopted, and the 1958 Convention is silent on the matter. The 1982 Convention has no provision either on length of baselines for coastal Archipelagos. As for the oceanic Archipelagos constituting the national territory of a State, the length of straight baselines must not normally exceed 100 nautical miles. However, up to 3 per cent of the total number of baselines for any Archipelago may reach a maximum of 125 nautical miles.[38]

9.5.   Consolidation of title

The question of consolidation of title was touched upon in previous chapters, but its importance for the present study necessitates a separate treatment. This Section will, therefore, examine the legal nature of the doctrine and the requirements for a consolidation of title to materialize.

Nature of consolidation of title

As indicated briefly in Chapter 5, the concepts of historic title and of consolidation of title are closely related. Both invoke history as a basis for the acquisition of soverignty over land or sea areas. The difference lies in the way in which history is invoked. In the case of *historic title*, history is relied upon as the sole basis for the claim of sovereignty. When such a claim applies to sea areas, an historical title represents an exception to the standard rules for the acquisition of maritime sovereignty and imports an exceptional burden of proof on the claimant State.

In the case of the *consolidation of title*, history is invoked only as a complementary and subsidiary basis, to solidify or consolidate the title resulting from the primary or main basis. Where the claim is to a sea area, the main basis might be a bilateral treaty as in the *Grisbadarna Case* of 1909, or it could be the straight baseline system, as in the *Fisheries Case* of 1951. The role of history being one of consolidation only and not the main basis for the claim, consolidation of title does not constitute an exception to standard rules for the acquisition of maritime sovereignty and does not import any special burden of proof on the claimant State.

The *Grisbadarna Case* is probably the first concrete and clear application of the doctrine of consolidation of title by the passage of time. The case involved a maritime boundary dispute between Norway and Sweden.[39] The Permanent Court of Arbitration had to decide whether the boundary line had been fully established by the boundary treaty of 1661 and, if not, where the line should be placed. The arbitral tribunal concluded that the treaty had not fully established the boundary line, that the probable intent of the treaty and accompanying map was not to continue the boundary along the median line beyond the islands and islets,[40] and that the unfixed part of the boundary should follow a line perpendicular to the general direction of the coast.[41] The boundary line was thus established so as to pass midway between the Grisbadarna banks, being awarded to Sweden, and Skjottegrunde, allotted to Norway. It follows that the main basis of the decision was the boundary treaty of 1661, interpreted in light of 'the notions of law existing at the time of the conclusion of the treaty'.[42]

Having found the main basis for awarding the Grisbadarna banks to Sweden, the arbitral tribunal invoked history as an additional and complementary basis to justify Sweden's sovereignty over the banks. The historical evidence accepted by the tribunal to support Sweden's claim pertained to lobster fishing and the installation of navigational aids. On lobster fishing, the Court found that 'lobster fishing in the shoals of Grisbadarna has been carried on for a much longer time, to a much larger extent, and by much larger number of fishers by the subjects of Sweden than by the subjects of Norway'.[43] Relating to navigational aids, the evidence showed that 'Sweden has performed various acts on the Grisbadarna region, especially of late, owing to her conviction that these regions were Swedish, as, for instance, the placing of beacons, the measurement of the sea, and the installation of a light-boat, being acts which involved considerable expense and in doing which she not only thought that she was exercising her right but even more that she was performing her duty'.[44]

The tribunal attached special importance to the historic lobster fishing rights of the local population.

> It is a settled principle of the law of nations that a state of things which actually exists and has existed for a long time should be changed as little as possible; and this rule is specially applicable in a case of private interests which, if once neglected, can not be effectively safeguarded by any manner of sacrifice on the part of the Government of which the interested parties are subjects.[45]

The 'settled principle' referred to by the tribunal is, of course, that of *quieta non movere*. Explaining that the main object of this principle was to protect the stability of territorial situations, Professor Charles De Visscher writes:

> It is for these situations, especially, that arbitral decisions have sanctioned the principle *quieta non movere*, as much out of consideration for the importance of these situations in themselves in the relations of States as for the political gravity of disputes concerning them. This consolidation, which may have practical importance for territories not yet finally organized under a State regime as well as for certain *stretches of sealike bays*, is not subject to the conditions specifically required in other modes of acquiring territory.[46]

In a footnote, De Visscher quotes from the *Grisbadarna Case* and states that 'this consideration had its part in the judgment of the Permanent Court of International Justice in the *Eastern Greenland Case*'.[47]

In the *Fisheries Case*, history was relied upon in two ways. First, Norway invoked history to justify its straight baseline system as a special application of the low-water mark rule. Second, it invoked historic rights as additional probative value to show the validity of a particular baseline. These were not equivalent to claims of historic waters, as explained by Professor Charles De Visscher, but they were claims of consolidation of title.[48] The Court accepted Norway's arguments and found, with respect to the straight baseline system, that it had been applied consistently for some 80 years[49] and 'that even before the dispute arose, this method had been *consolidated* by a constant and sufficiently long practice, in the face of which the attitude of governments bears witness to the fact that they did not consider it to be contrary to international law'.[50] With respect to the waters of the Lopphavet, the Court held that the historic fishing and hunting rights of the land inhabitants justified the line across that basin by way of consolidation of title.

Requirements of consolidation of title

Since the doctrine of consolidation of title does not serve as the main ground for the acquisition of sovereignty, the requirements are not as stringent as for the acquisition of an historic title.[51] However, virtually the same elements are present and the same questions are raised. Consequently, somewhat as in the previous treatment of historic title in Chapter 6, the following will be addressed: (i) exercise of State authority; (ii) long usage or passage of time; (iii) general toleration; (iv) legal effect of protest; (v) vital interests of the coastal State; and (vi) burden of proof.

Exercise of State authority

As was seen in the *Grisbadarna Case*, this first element seems to have been given extensive interpretation. The installation of navigational aids was, of course, done by State authorities, but there was no evidence in the decision that the lobster fishing on the Grisbadarna banks had taken place on the basis of a special licence granted by the Government. In addition, the lobster fishing by the Swedish fishermen had not been to the exclusion of Norwegian fishermen. In the *Fisheries Case*, however, the Norwegian Government had granted a number of licences which provided for the exclusive right to fish and hunt whales in the waters of the Lopphavet. In the same way for the straight baseline system itself, the government authorities had provided for the system in its legislation and had enforced it.

It follows from the above examples that the exercise of State authority normally involves legislation or regulations, but not necessarily so. Indeed, the arbitral tribunal in the *Grisbadarna Case* specified that the *quieta non movere* rule 'is especially applicable in a case of private interests' of citizens of the claimant State.[52]

Long usage or passage of time

Long usage is by far the most important requirement for an historical consolidation of title. It is really its foundation, as emphasized by Professor De Visscher: 'Proven long use, which is its (consolidation) foundation, merely represents a complex of interests and relations which in themselves have the effect of attaching a *territory or an expanse of sea* to a given State'.[53] The learned author is careful to add, however, that there is no requirement for the passage of a fixed term which, he says, is unknown to international law.[54]

The period of time will, of course, vary from one situation to another. There was no specific period mentioned in the *Grisbadarna* Case but, presumably, the lobster fishing began at about the time of the boundary treaty in 1661. In the *Fisheries Case*, the first fishing and whale hunting licence was granted at the end of the 17th century. As for the straight baseline system itself, it had first been adopted by Norway about 80 years prior to the dispute. The period of time for the consolidation therefore varied in those instances from 80 to about 250 years.

General toleration

Acquiescence is not required here in the same way as it is for a claim of historic title properly so-called. All that is necessary is that there be 'a sufficiently prolonged *absence of opposition* either, in the case of land, on the part of States interested in disputing possession or, in maritime waters, on the part of the generality of States'.[55]

In the *Grisbadarna Case*, the lobster fishing by Swedish fishermen had obviously been acquiesced in by Norway, whereas there was merely the absence of protest toward Sweden's placing of beacons and a light-boat.[56] In the *Fisheries Case*, the Court held that 'the *general toleration* of foreign States with regard to the Norwegian practice is an unchallenged fact'.[57] It so held, in spite of the United Kingdom's formal protest in 1933, because its silence for a period of more than 60 years after the first enactment of Norway's straight baseline system represented too long an abstention.[58]

Legal effect of protest

The question of what constitutes an effective protest and its legal effect was discussed in Chapter 6 in connection with the requirements of an historical title, and what was said there applies here as well. It should perhaps be added that, since a consolidation of title is not a root of title *per se* but serves only to solidify the real root, it is only logical that a stronger protest should be necessary to prevent the consolidation from materializing. This, in effect, is simply the converse of the earlier related proposition that a general toleration is sufficient and no real acquiescence is necessary. This seems to have been the view of the International Court in the *Fisheries Case*, when it held that the 'formal and definite protest'[59] by the United Kingdom was not sufficient to prevent a consolidation of Norway's baseline system.

Vital interests of the coastal State

By its very flexible nature, the doctrine of consolidation of title easily encompasses the vital interests of the coastal State and its inhabitants, as constituting an admissible factor in addition to those just reviewed. In effect, certain vital interests might be considered as equitable in nature, in the sense that it would be unfair to the coastal State not to take them into account. In the *Grisbadarna Case*, the tribunal not only took the private interests of the local fishermen into account but expressly stated that the rule *quieta non movere* 'is especially applicable in a case of private interests'.[60]

In the establishment of its straight baseline system by its 1935 decree, Norway invoked in the preamble the following vital and related interests: 'well-established national titles of right', 'the geographical conditions prevailing on the Norwegian coasts' and 'the safeguard of the *vital interests* of the inhabitants of the northernmost parts of the country'.[61] At the time of its 1869 Decree in which it had fixed a fishery limit by drawing a straight line of more than 10 sea miles, Norway stated in its Note to France that '*local conditions* and considerations of what is practicable and *equitable* should be decisive in specific cases'.[62] The Court accepted Norway's arguments, by giving considerable weight to the 'vital needs of the population'[63] of the Lopphavet region and holding that 'the method of straight baselines . . . was imposed by the peculiar geography of the Norwegian coast'.[64]

Burden of proof of consolidation of title

Since this question was discussed earlier in Chapter 6 when dealing with the requirements of an historic title, suffice it to state here that, because a consolidation of title does not constitute an exception to the rules of international law for the acquisition of maritime sovereignty, there is no special burden of proof imposed upon the claimant State.

The Court did not deal expressly with this question in the *Fisheries Case*, but it was satisfied that Norway's claim was one of consolidation of title and it must also have been satisfied that there was no special burden of proof on Norway.

9.6.   Summary

1. According to the *Fisheries Case*, the geography required for the applicability of straight baselines is that the coast be deeply indented or 'bordered by an Archipelago such as the Norwegian skjaergaard'. The Territorial Sea Convention of 1958 and the Law of the Sea Convention of 1982 require either deep indentations or a 'fringe of islands along the coast in its immediate vicinity'. Both Conventions contain a customary law saving clause which provides that matters not regulated in the Conven-

tions continue to be governed by rules of general international law.

2. According to the *Fisheries Case*, the straight baseline system represents 'the application of general international law to a specific case', in that the baselines constitute a special implementation of the low-water mark rule. The 1958 and 1982 Conventions provide for the straight baseline system as an exception to the low-water mark rule.

3. The Court in the *Fisheries Case* formulated three criteria or guidelines, which may be adapted to diverse situations in drawing straight baselines. All three criteria were incorporated without change in the 1958 and 1982 Conventions. The first two criteria are compulsory and the third is optional. They may be summarized as follows:

   (1) Straight baselines must not depart to any appreciable extent from the general direction of the coast, in that the deviation must not constitute a distortion of the general direction. To judge whether a baseline constitutes a distortion, a global view of the whole coast should be had on a small-scale map.

   (2) The enclosed sea areas must be sufficiently closely linked to the land domain to be subject to the regime of internal waters. The overall ratio of sea to land areas within the Norwegian Archipelago was 3.5 to 1. Under the *Fisheries Case*, the enclosed waters are strictly internal and no right of innocent passage exists. Under the 1958 and 1982 Conventions, the newly enclosed waters are internal but subject to the right of innocent passage if they had previously been considered as part of the territorial sea or of the high seas.

   (3) In establishing particular baselines, account may be taken of certain economic interests peculiar to the regions concerned, the reality and importance of which are clearly evidenced by long usage. Such evidence adds probative value to the validity of the baselines in question.

4. Neither the *Fisheries Case* nor the 1958 or 1982 Conventions fix any maximum length for straight baselines enclosing coastal Archipelagos. The coastal State can best appraise the local conditions dictating the precise position of baselines.

5. A consolidation of title is not meant to constitute a sole basis for a claim of maritime sovereignty; this is the function of an historic title. In a consolidation of title, history is invoked only as a subsidiary basis in order to consolidate the title resulting from a

primary basis such as the straight baseline system. A consolidation of title does not constitute an exception to standard rules for the acquisition of maritime sovereignty.

6. The requirements for a consolidation of title are basically of the same nature as those for an historic title, but they are not as stringent. The requirements are the following:

    (1) exercise of State authority, which might be limited to legislation or regulations;

    (2) long usage or the passage of a long period of time, the length of the period depending on the circumstances; and

    (3) a general toleration or absence of formal protest by foreign States, particularly those clearly affected by the claim.

7. An effective protest should prevent a consolidation of title from materializing, but it would have to be somewhat stronger than in a case of historic title.

8. In a claim of consolidation of title, the coastal State may legitimately invoke its vital interests, such as the peculiar geographical conditions of its coast and its duty to protect the economic interests or vital needs of its local inhabitants.

9. Since a claim of consolidation of title does not constitute an exception to the standard rules for the acquisition of maritime sovereignty, there is no special burden of proof on the claimant State.

10

State practice on the use of straight baselines

Since the approval of straight baselines by the International Court of Justice in the *Fisheries Case* of 1951, their use has become increasingly common in the practice of States for the delimitation of their territorial sea. This increased practice applies to all types of archipelagos, whether or not they fit into the two categories provided for in the 1982 Convention. Consequently, this chapter will review State practice in general and that of Arctic States in particular.

10.1. State practice in general

In 1971, when Commander P. B. Beazley of the British Admiralty Hydrographic Service made his study on the use of baselines, he concluded that straight baselines were being used by 24 States.[65] The baselines varied in maximum length from 23 to 222 nautical miles. By 1985, 60 States had used the straight baseline system and 12 more had adopted enabling legislation. This means that a substantial majority of coastal and Archipelagic States have either actually established straight baselines in the measurement of their territorial sea or are preparing themselves to do so. Table 1 lists the 66 countries and the longest straight baseline established by each. For four of the States listed (Ecuador, The Soviet Union, Denmark and Norway), several maximum lengths are shown.

10.2. State practice in the Arctic

Norway

It was Norway in 1812 which adopted the first enabling legislation to measure its territorial sea from straight baselines between the outermost islands from its mainland. The Royal Decree of 1812 limited itself to saying that the territorial sea of one ordinary sea league would be measured 'from the island or islet farthest from the mainland',[67] but in its

Table 1. *Maximum straight baselines.*[66]

This table includes only the longest straight baselines properly
so-called and does not include closing lines across historic bays.

State	Length (in nautical miles)
1. Burma	222.3
2. Bangladesh	220
3. Vietnam	161.8
4. Philippines	140
5. Cape Verde	137
6. Ecuador (mainland)	136
7. Colombia	130.5
8. Fiji	125
9. Ecuador (Galapagos)	124
10. Indonesia	124
11. Solomon Islands	124
12. Madagascar	123
13. Malaysia (for delimitation of continental shelf)	123
14. Papua New Guinea	121
15. Guinea	120
16. Argentina/Uruguay (by agreement)	118
17. Soviet Union (Pacific Ocean)	103.8
18. Sao Tomé and Principe	101
19. Canada	99.5
20. Venezuela	98.9
21. Vanuatu	93
22. Iceland	92
23. Haiti	89
24. Mauritania	89
25. Denmark (Greenland)	75
26. Australia	71.6
27. Cuba	69
28. Chile	65
29. Denmark (Faeroes)	60.7
30. Mozambique	60.4
31. Soviet Union (Arctic Ocean)	60.1
32. Korea, Democratic People's Republic	60
33. Ethiopia	57.5

State	Length (in nautical miles)
34. Spain	50.5
35. Tunisia	44.7
36. Norway	44
37. Tanzania	44
38. United Kingdom	40.2
39. Mexico	39.4
40. Italy	37.5
41. France	34.5
42. Thailand	33.7
43. Cambodia	33
44. Poland	30
45. Sweden	30
46. Guinea-Bissau	29
47. Ireland	25.2
48. Morocco	24.5
49. Kenya	24
50. Turkey	23.5
51. New Zealand	23
52. Dominican Republic	22.7
53. Germany, Democratic Republic	22.7
54. Yugoslavia	22.5
55. Senegal	22
56. Denmark (mainland)	21.8
57. Germany, Federal Republic	21.5
58. Albania	21.2
59. Angola	20.3
60. Norway (Svalbard)	18.5
61. Norway (Jan Mayen)	18
62. Cameroon	17.5
63. South Africa	17.3
64. Gabon	16.5
65. Oman	12
66. Finland	8

States with enabling legislation

Brazil	Malta	Somalia
China	Sri Lanka	Sudan
Egypt	Mauritius	Syria
Iran	Saudi Arabia	Maldives

practice Norway consistently construed the Decree as enabling it to establish straight baselines between the outermost islands.[68] The subsequent decrees of 1869, 1881, 1889 and 1935 simply brought greater precision to the 1812 Decree and gradually provided for the establishment of straight baselines between all the islands and islets along the whole coast of Norway northward of 66° 28.8' N latitude. The International Court approved the baseline system in 1951, as well as all of the actual baselines established by Norway in its 1935 decree.[69] Three of those baselines were particularly objected to by the United Kingdom: the first across the Svaerholt basin of 38.6 miles long, the second across the Lopphavet basin of 44 miles running to a submerging rock, then a further 18 miles to another similar rock, and the third line across the Vestfjord which is 40 miles in length. In approving the line across the Lopphavet basin, the Court took into account the historic fishing and whale hunting rights of the local population.[70]

In 1955, Norway adopted a Decree drawing straight baselines around the Island of Jan Mayen and outlying islets.[71] The longest baseline established was 18 miles. In 1970, Norway delimited its territorial sea around part of Svalbard, 'from Verlegenhuken to Halvmaneova and around Bjornoya and Nopen' by drawing straight baselines.[72] It has not yet arrived at an agreement with the Soviet Union as to the delimitation of the continental shelf of Svalbard in the Barents Sea. In 1976, the exclusive economic zone adopted by Norway was, of course, measured from 'established baselines'.[73]

Denmark

In 1963, Denmark delimited a territorial sea of three nautical miles off the coast of Greenland from the low-water mark or straight lines between the points specified in the Executive Order.[74] The length of the straight baselines varied from a few miles to 75 nautical miles. Those baselines served also to delimit the continental shelf between Greenland and the Canadian Arctic Islands in 1973.

In 1966, Denmark delimited its territorial sea for its mainland in the same way as it had for Greenland, that is either from the low-water mark or straight baselines between specified points. The existing right of passage in the normal routes of the three belts and Sound continues to apply.[75] In 1976, Denmark established straight baselines around the Faeroe Islands, one of the baselines being 60.7 miles in length. The exclusive fishing zone, adopted by Denmark in 1976, is also measured from straight baselines wherever they exist.[76]

United States

The United States does not provide for the use of straight baselines in its legislation and has never made use of such baselines. Indeed, its traditional position has been against the use of straight baselines to delimit territorial waters of Archipelagos, whether they be coastal or oceanic. This position, however, allows for exceptions in relation to certain Archipelagos and seems to have undergone a conditional change with respect to oceanic Archipelagos.

In relation to coastal Archipelagos, it accepts the use of straight baselines 'in relatively few areas of the world, such as along the highly irregular and fragmented coasts of Yugoslavia, Norway and Southern Chile'.[77] But, since it has not ratified the Territorial Sea Convention of 1958 nor signed the Law of the Sea Convention of 1982, it is difficult to say if it accepts the provisions of those conventions on coastal Archipelagos. Certainly, it has not drawn straight baselines along its own coasts, which it could well have done along the coast of Alaska for instance.

With respect to oceanic Archipelagos, the United States has traditionally taken the view that the use of straight baselines to form a perimeter around a group of islands was 'no more justified than a corresponding line along the mainland'.[78] It adopted this position in 1947 for the Pacific Islands, formerly under Japanese mandate and transferred to the U.S. trusteeship after World War II, and has objected to other countries using straight baselines for similar Archipelagos. In 1951, it protested against Ecuador's assertion of a claim to a single belt of territorial waters around the entire Colon Archipelago as being contrary to international law.[79] In 1957, it sent a Note of protest to Indonesia and, in 1958, to the Philippines when those countries claimed the waters within their respective waters as internal waters.[80] The United States reiterated these protests in 1960 at the time of the Second Law of the Sea Conference, when the head of its delegation wrote that 'we do not recognize the validity of this extensive and unilateral Archipelago theory'.[81] In 1964, the United States maintained the same position with respect to Hawaii when Secretary of State Rusk replied to the Attorney General that it was 'the Department's position that each of the islands of the Hawaiian Archipelago had its own territorial sea, three miles in breadth measured from low-water mark on the coast of the island'.[82]

This traditional position has undergone a conditional change during the Third Law of the Sea Conference, in that the United States now seems to recognize the possibility for Archipelagic States to draw straight baselines around their Archipelago on certain conditions. Those conditions were incorporated in the 1982 Convention and are basically three in number: first, that the water/land ratio fall between 1 to 1 and 9 to 1; second, that the baselines not exceed 100 nautical miles in length except for up to 3 per cent which may reach 125 nautical miles; and third, that routes normally used for international navigation and overflight (or suitable substitutes) be maintained.[83] The right of 'sea lanes passage' applicable in these routes permits the passage of all ships, including the submerged transit of submarines.

The conditional change in the United States' position applies only to Archipelagic States, that is 'constituted wholly by one or more Archipelagos'.[84] This was made clear in a 1978 letter from the Department of State, replying to a question by Senator Clairborne Pell in the following terms:

> Certain states, which claim they are legally or geographically Archipelagic States, have sought international recognition that the waters within the baselines connecting the outermost points of their outermost islands are internal or Archipelagic waters and that the territorial sea and fisheries zone should extend seaward of those baselines. *The United States' position is that under international law the territorial sea and other maritime jurisdictions are to be drawn around each island.*[85]

To summarize the present position of the United States relating to straight baselines, it may be stated firstly, that it recognizes their applicability to the highly irregular and fragmented coasts of Yugoslavia, Norway and Southern Chile, and secondly, that it also recognizes their applicability to the islands of Archipelagic States, subject to certain conditions of water/land ratio, lengths of baselines and rights of passage.

Soviet Union

The legislation of the Soviet Union provides that its 12-mile territorial sea is to be measured 'from the lowest ebb-tide both on the mainland and on islands which belong to the U.S.S.R., or from *straight baselines* joining appropriate points'.[86] This provision of the 1982 Law on the State Boundary is slightly different from a previous one adopted in 1971 which followed the wording of the 1958 Convention and made straight baselines applicable 'in those localities where the coastline is deeply indented and cut into or if there is a fringe of islands along the immediate vicinity'.[87] In omitting these qualifying words as to the geographic requirements for the applicability of straight baselines, the Soviet Union gave itself considerable latitude as to where such lines may be used.

In February 1984, the Council of Ministers adopted a Decree establishing straight baselines along the east side of the Kamchatka Peninsula in the Bering Sea, the longest ones varing from 44 to 103.8 miles. This was followed by another Decree on January 15, 1985, adopting straight baselines for the northern coast in the Arctic ocean, as well as around three Archipelagos: Novaya Zemlya (New Land), Severnaya Zemlya (North Land) and Novosibirskiye Ostrova (New Siberian Islands).[88] The Archipelago of Novaya Zemlya, separating Barents Sea from Kara Sea, extends almost at right angles from the mainland coast to which it is joined by straight baselines across three straits, the main one being Kara Gates Strait. This strait is enclosed by straight baselines 29 miles long at its western entrance and one of 32 miles at its eastern entrance.

The Severnaya Zemlya Archipelago, which projects northwesterly deep into the Arctic Ocean and separates Kara Sea from Laptev Sea, is linked to the mainland by straight baselines across four straits (see Figure 14). The main strait enclosed and which constitutes a key section of the Northeast Passage is Vilkitsky Strait(s), the entrances of which are divided by the presence of islands thus forming at least two straits. The western entrance is closed by a straight baseline of 60.1 miles drawn in a southwesterly direction from Bolshevik Island to the north of the strait to a group of islands north of the mainland. The main baseline at the eastern entrance of the Vilkitskii Straits is 44 miles long.

The third Archipelago involved, separating Laptev Sea from the East Siberian Sea, is that of the New Siberian Islands. Straight baselines are established around the western part of the Archipelago, enclosing the Sannikov, Eterikan and Dimitrii Laptev Straits. The straight baselines across the main strait, Sannikov, measure 31.3 miles at the western entrance and 43.4 miles at the eastern one.

A fourth and northernmost Archipelago of the Soviet Union, Zemlya Frantsa Iosifa (Franz Joseph Land), situated in the Barents Sea and close to Svalbard, has not been enclosed by straight baselines. Negotiations between Norway and the Soviet Union on the maritime boundary between the two Archipelagos have not yet resulted in a delimitation agreement.

The 1985 Decree specifies that the territorial waters, the exclusive economic zone and the continental shelf are to be measured from the newly established straight baselines. This makes it clear that all the enclosed areas, including the straits just described, become internal waters of the Soviet Union. Presumably, however, these newly enclosed waters would be subject to the right of innocent passage, since they were previously either territorial waters or high seas (or exclusive economic

Fig. 14. Straight baselines of Arctic Archipelagos of U.S.S.R. *Source*: Prepared for the author by John E. Cooper, Technical Advisor to Dept of External Affairs, Ottawa.

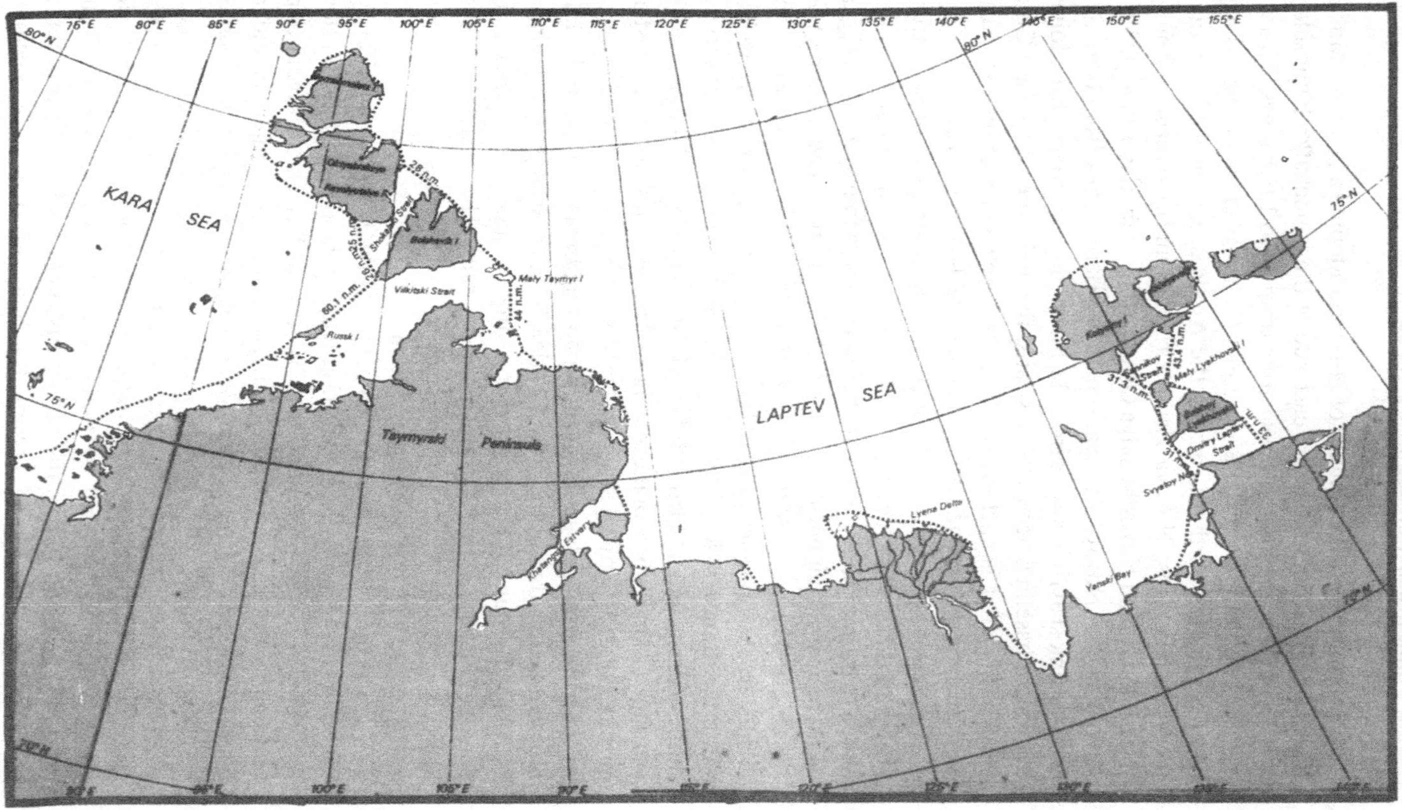

zone) and the Soviet Union is a Party to the 1958 Territorial Sea Convention.

Canada

In 1964, Canada provided for the use of straight baselines in its Territorial Sea and Fishing Zones Act. This Act empowers the Executive to issue lists of geographical coordinates of points from which baselines may be determined and 'subject to any exceptions in the list for the use of the low-water line along the coast as the baseline between given points, baselines are *straight baselines* joining the consecutive geographical coordinates of points so listed'.[89] Pursuant to this legislation, Canada has promulgated two series of straight baselines: one, in 1967, for the coasts of Labrador, Newfoundland and Nova Scotia and the other, in 1969, along the west coast of Queen Charlotte Islands and Vancouver Island.[90] Except for the lines across certain historic bays along the coast of Newfoundland (Notre Dame, 49 miles; Placentia, 45 miles; and Bonavista, 31 miles), the longest straight baseline is 29 miles.

In the Arctic, Canada began by establishing delimitation lines outside its Archipelago for various purposes. In 1970, it adopted a 100-mile pollution prevention zone; in 1973, it agreed on a continental shelf delimitation line with Denmark; and, in 1977, it adopted a 200-mile exclusive fishing zone.[91] Finally, on September 10, 1985, Canada established straight baselines around the perimeter of the Canadian Arctic Archipelago (see Figure 15). These baselines, which came into force on January 1, 1986,[92] defined the 'outer limit of Canada's historic internal waters'.[93] Accordingly, the status of internal waters would result from a historic title rather than from the enclosure of the Archipelago by straight baselines. However, if no historic title can be proved, which is the conclusion arrived at in Chapter 8, the status of internal waters must result from the drawing of straight baselines. The enclosed waters would then be subject to the right of innocent passage if the lines are drawn under the Territorial Sea Convention of 1958, but would not be if they are drawn under customary law by virtue of the *Fisheries Case* decision of 1951. Since Canada is not a Party to the 1958 Convention, it may rely on the *Fisheries Case*. This possiblity is discussed in Chapter 14 on the legal status of the Northwest Passage.

10.3. Summary

1. The great majority of coastal States have either actually used straight baselines in the measurement of their territorial sea or have enabling legislation to do so.

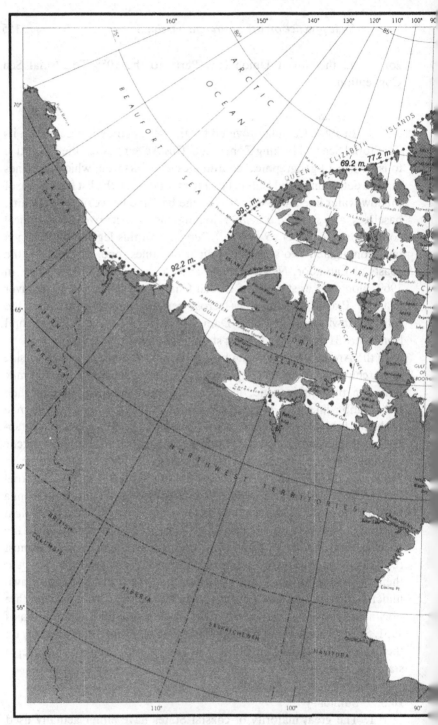

Fig. 15. Straight baselines of Canadian Arctic Archipelago. *Source*:
Prepared for the author by John E. Cooper, Technical Advisor to Dept
of External Affairs, Ottawa.

156

2. The longest baseline for each State varies from 8 to 222.3 nautical miles.
3. Four of the five Arctic States (Norway, Denmark, the Soviet Union and Canada) have made use of straight baselines to measure their territorial sea. The United States has no enabling legislation and has not used straight baselines in measuring its own territorial sea. It does recognize, however, the applicability of the straight baseline system to highly irregular and fragmented coasts such as those of Yugoslavia, Norway and Southern Chile. The United States also recognizes the applicability of straight baselines to the islands of Archipelagic States under the conditions laid down in the Law of the Sea Convention of 1982.

11

Straight baselines applied to the Canadian Arctic Archipelago

The possibility of enclosing the Canadian Arctic Archipelago with straight baselines has been discussed by a number of writers since the *Fisheries Case* of 1951, particularly after the *Manhattan* crossing of the Northwest Passage in 1969. Their writings have already been reviewed by the present writer and suffice it to recall here that, subject to a few nuances, they all arrived at the conclusion that those waters could be enclosed in a way similar to those of the Norwegian Archipelago.[94] In spite of the virtual unanimity in the conclusion, the reasons varied somewhat from one writer to another, particularly as to the precise legal nature of the Canadian Archipelago, the use which could be made of history and the consequence which straight baselines would have on any right of passage that might now exist. In light of the applicable law and the practice of States already discussed, this chapter will examine: (1) the geography of the Canadian Arctic Archipelago; (2) straight baselines for the Archipelago; and (3) the consolidation of title to certain waters enclosed by straight baselines.

11.1. The geography of the Canadian Arctic Archipelago

In the *Fisheries Case* of 1951, the International Court concluded that 'the method of straight baselines, established in the Norwegian system, was imposed by the peculiar geography of the Norwegian coast'.[95] The question arises here whether the geography of the northern coast of Canada is of a similarly peculiar nature so as to warrant the method of straight baselines for the delimitation of its territorial waters. What follows is a description of the basic geographic realities which must be kept in mind in trying to determine if the criteria for the applicability of the straight baseline system are fulfilled.

All of the Canadian Arctic Archipelago except for the southern tip of Baffin Island, lies north of the Arctic circle, and constitutes the northern coastal zone of Canada (see Figure 15). The base of this triangular-shaped Archipelago stretches some 3000 kilometres along the mainland coast and its apex, the tip of Ellesmere Island, is less than 900 kilometres from the geographic North Pole. The Archipelago is one of the largest in the world and consists of a labyrinth of islands and headlands of various sizes and shapes. There are 73 major islands, of more than 50 square miles in area, and some 18,114 smaller ones. The very large islands are Baffin, Devon and Ellesmere on the east side of the Archipelago and Victoria, Banks and Melville on the west. Virtually all of the land formations are mountainous in character.

The western part of the mainland coast is broken by large indentations in the form of bays and gulfs, and the eastern section is deeply penetrated by a huge inland sea (Hudson Bay) and smaller bays and basins. Nearly all of these bodies of water are seeded with countless islands, rocks and reefs. Consequently, the coast of the mainland does not constitute at all a clear dividing line between land and sea, as it does in most other countries. In fact, the coast reaches northward as far as an east–west waterway (Parry Channel) crossing the middle of the Archipelago; it does so by way of a long northern projection (Boothia Peninsula), barely broken by an extremely narrow strait (Bellot Strait) to form Somerset Island to the north.

To the north of Parry Channel, the Queen Elizabeth Islands group, constituted of large and small islands of various shapes, virtually all of them deeply indented, is interspersed with bodies of water equally varied in size and shape. This northern section of the Archipelago is linked with the southern one by a string of five islands lying in a zigzag fashion across west Barrow Strait in Parry Channel, thus forming inter-island passages varying from 8 to 15.5 miles. The islands and peninsulas of the whole Archipelago are fused together by ice formations most of the year, to the point where ice and land areas often become indistinguishable. The Archipelago then transforms itself into an immense rampart, protecting the continental part of Canada from the polar ice of the Arctic Ocean and constituting in effect the outer coast of the country. The inhabitants of this barren coastal zone derive their livelihood from hunting and fishing, traversing the ice and land indifferently by dog sled or snowmobile. *Such are the realities which must be borne in mind in determining the applicability of the straight baseline system.*

In light of the foregoing, the specific question to be aswered is whether the northern coast of Canada 'is bordered by an Archipelago such as the "skjaergaard" ', as the International Court put it,[96] or constitutes a 'fringe of islands along the coast in its immediate vicinity', in the words of the Conventions.[97] It is suggested that there are two elements in this prerequisite: the closeness of the Archipelago to the coast and the cohesiveness of the Archipelago itself. As for the *closeness* to the coast, there can be no question that this element is present, since not only are most of the islands forming the base of the Archipelago located very close to the coast, but the coast itself, through its central peninsula, advances into the very core of the Archipelago. Regardless of whether one applies the term 'bordered' or the expression 'immediate vicinity', this first element of the criterion is fully met.

As for the *cohesiveness* of the Archipelago itself, there exists a general interpenetration of land formations and sea areas, and this close relationship is reinforced by the presence of ice most of the year. The geographic unity of the Archipelago is further assured by the string of closely spaced islands across Parry Channel, linking the northern section with the southern one and forming a single unit. Admittedly, Parry Channel cannot be fully compared with the narrow Norwegian Indrelia which the Court held was 'not a strait at all, but rather a navigational route prepared as such by means of artificial aids to navigation provided by Norway'.[98] However, in spite of the considerable width of Parry Channel, a global view shows that it does not unduly disrupt the general unity of the Archipelago. The overall result is that the Canadian Arctic Archipelago presents very special geographic characteristics, making it absolutely impossible to follow the sinuosities of the coast or of the islands in the measurement of territorial waters and rendering the use of straight baselines necessary.

The Archipelago might not constitute a 'fringe of islands along the coast', if the Conventions are interpreted literally, but such an interpretation would not be in accord with the practice of States. This practice indicates that States have been either ignoring the precise wording of the provisions of the Conventions or interpreting them as a mere codification of the criteria laid down in the *Fisheries Case* which, indeed, they were intended to be. Professor O'Connell lists some 18 coastal Archipelagos where straight baselines were used and which constitute very doubtful 'fringes' of islands.[99] And, as asked by the learned author, 'how many islands make a "fringe", and what must be their relationship one to another and to the mainland?'[100] This question, he says, can only be answered by looking at the criteria to be met in the actual drawing of the baselines, such as the general direction of the coast. These criteria will now be applied.

11.2.   Straight baselines of the Canadian Arctic Archipelago

The straight baseline system being applicable, the task now is to draw the baselines along the Archipelago in such a way that they will follow the general direction of the coast and that the sea areas will be sufficiently linked to land. The justification of some of the lines may be reinforced by reference to regional economic interests evidenced by long usage. As shown on Figure 15, baselines begin at the 141st meridian of longitude, proceed in a general easterly direction along the continental coast of Canada in the Beaufort Sea as far as Baillie Islands off Cape Bathurst at the entrance of Amundsen Gulf, continue in a northeasterly direction across the Gulf and along the west side of Banks Island, across M'Clure Strait and along the perimeter of the Queen Elizabeth Islands as far as the most easterly point of Ellesmere Island in the Lincoln Sea. The baselines then proceed in a general southerly direction along the perimeter of the Islands to Lancaster Sound, across the Sound and along Bylot and Baffin Islands as far as Resolution Island at the entrance of Hudson Strait where they join the straight line across that Strait established in 1937. Although there is no maximum length for straight baselines of coastal Archipelagos, it is of interest to note that the 145 baselines vary from a few hundred yards to 99.5 nautical miles and result in an average length of less than 17 nautical miles.

General direction of the coast

Considering the triangular shape of the Archipelago, the only possible general direction which straight baselines can follow, after reaching the entrance of Amundsen Gulf, is that of the outer line of the Archipelago itself. The geographic realities are such that it is absolutely impossible to follow the general easterly direction of the coast. In addition, that general easterly direction ends with the Boothia Peninsula which projects at right angles in a northerly direction in effect as far as Parry Channel, and after that there is no general direction of the coast. On the contrary, there is immediately to the east of Boothia another northerly projection (Melville Peninsula) reaching the underside of the western extremity of Baffin Island which constitutes the eastern limit of the Archipelago.

Taking into account these two important northerly projections of the coast in the middle of the Archipelago, it would be technically correct to say that those baselines, which do not follow the general easterly direction of the coast, follow its general northerly direction. But this would be stretching the facts somewhat to fit the law, whereas the opposite should normally be done. As was once properly suggested by the great American judge and jurist, Benjamin Cardozo, if the law does not quite fit the facts the judge should adapt the law to the facts and not do the reverse.

The International Court did adapt the law to the facts in 1951 when it realized that the low-water mark rule could not be applied to the Norwegian Archipelago and approved the straight baseline system. Similarly, realizing that the first guideline for the application of the system is not completely appropriate for the Canadian Archipelago, being shaped rather differently than the Norwegian skjaergaard although very similar in character, an international tribunal should adapt the guideline to the special geographic reality. More specifically, it should consider that what really constitutes the Canadian coastline is the outer line of the Archipelago – in the same way that the International Court considered that 'what really constitutes the Norwegian coast line is the outer line of the skjaergaard'[101] – and permit the baselines to follow the outer line or general direction of the Archipelago.[102]

Close link between land and sea
The International Court judged it of fundamental importance that, as a rule, the sea areas be sufficiently closely linked to the land domain to be subject to the regime of internal waters. However, here again, the Court applied this guideline liberally to at least two areas of the Norwegian Archipelago, the Lopphavet and the Vestfjorden. In a similar way, the flexibility of this guideline should permit the enclosure of the waters of Amundsen Gulf and Parry Channel.

At least three reasons militate in favour of such a flexibility. First, the sea to land ratio in the Canadian Arctic Archipelago is 0.822 to 1,[103] considerably better than the 3.5 to 1 ratio for the Norwegian Archipelago. Second, the quasi permanent presence of ice over the enclosed waters, used like land for travels by dog sled and snowmobile, and even for human habitation by the Inuit during their winter hunting trips, bolsters the physical unity between the land and the sea. Third, the innocent passage of foreign ships should, and presumably would, be permitted by Canada;[104] if so, and the exclusion of foreign ships being the main reason for the close link requirement, the importance of this requirement is considerably lessened and its application should be correspondingly liberalized.

Regional economic interests evidenced by long usage
Although the straight baselines would be justified under the two compulsory guidelines just reviewed, Canada is in a position to invoke certain economic interests peculiar to some regions, the reality and

importance of which are clearly evidenced by long usage. These interests may be relied upon to reinforce the validity of certain baselines, particularly the 51-mile line across Lancaster Sound at the eastern end of Parry Channel and the 92-mile line across Amundsen Gulf on the west side of the Archipelago. Considering that these two bodies of water are located at either end of the most likely route for the future shipping envisaged for the Northwest Passage, it becomes particularly important that the baselines across these water areas be fully justified. It will be recalled that the interests taken into account by the Court in the *Fisheries Case* were the traditional rights reserved for the local inhabitants over fishing grounds, such rights being founded on the vital needs of that population.[105]

In the Canadian Arctic it has now been established that the Inuit have been fishing, hunting and trapping in the waters and on the sea ice of most of the Archipelago (see Figure 16). The government-sponsored study on Inuit land use and occupancy completed in 1976[106] reveals that their traditional sea ice use has covered all the waters of the central and eastern Arctic, as well as those of the western Arctic as far west as Canada's boundary in the Beaufort Sea and in a northerly direction up to M'Clure Strait and Viscount Melville Sound. This traditional hunting and trapping on the sea ice is still vital today to the Inuit economy.

Numerous communities rely on marine-related fur exports for the majority of earned income. For example, in 1969 the western Arctic communities of Holman, Paulatuk and Sachs Harbour acquired 53 per cent, 78 per cent and 54 per cent respectively of their earned incomes from the seals, white fox and polar bears.[107] During the 1970s the value of country foods harvested by the Inuit in the western Arctic averaged nearly $10 million annually, almost equivalent to their total yearly cash incomes.[108] In the eastern Arctic, meanwhile, the communities of Resolute and Arctic Bay acquired 60 per cent of their incomes from animal harvests in 1978–9, while Grise Fiord and Pond Inlet showed an even greater reliance of 79 per cent and 83 per cent respectively.[109] The average yearly value of animal harvests to each hunter is $2,792 in Resolute, $5,165 in Arctic Bay, $5,788 in Pond Inlet and $9,385 in Grise Fiord.[110]

In addition to providing a vital income, the marine mammals insure a high protein content in the diet of the Inuit which is essential to the maintenance of their health and energy in the cold and rigorous climate of the Arctic. As noted by anthropologists, subsistence foods with their high oil and fat content are particularly vital for the sick and the elderly.[111] The importance of local natural foods for the health of the native population

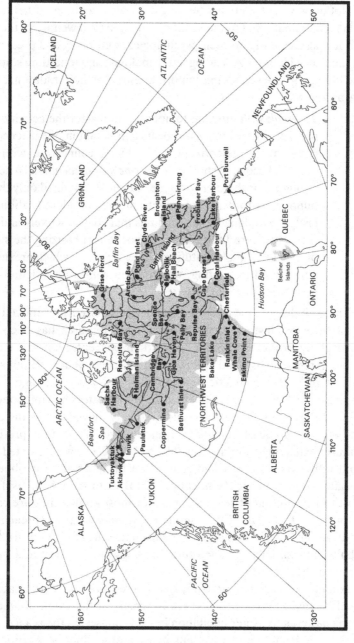

Fig. 16. Extent of ice use by the Inuit. Prepared for the author by the Canadian Hydrographic Service, Dept of Fisheries and Oceans, Ottawa. *Source: Report on Inuit Land Use and Occupancy Project, 1976, Vol. 3, 153.*

165

was also recognized in a study made in 1980 by the Science Advisory Board of the Northwest Territories.[112]

Not only is the traditional hunting of whales, seals and other marine-related mammals vital to the economic and physical welfare of the Inuit, but it is also essential to their psychological well-being and preservation of their own identity. A leading anthropologist and long-time student of the Inuit people describes this important aspect of their traditional way of life as follows:

> *To break with hunting*, to not seek to master the activity whose practice assured one's forebears a millennium of existence, *is* to deny in one essential way the living connection with one's ancestral roots; in short, *to deny one's Inuit identity.* To not seek to obtain, or share or consume real meat, obtained only through hunting, is to deny important emblematic identification. And lastly, to attenuate the social ties that so distinctively and effectively structure Inuit society by not re-emphasizing the salience of those essential institutions and values . . . which are inextricably linked to hunting, would again constitute a denial of full membership in that society.[113]

These vital rights and interests have been exercised and enjoyed by the Inuit since pre-historic times. Considerable research has been done, particularly in the last decade, as to their concentration and mode of living. Archaeologists and anthropologists believe that the Inuit arrived in Canada's western Arctic between 4000 and 4500 years ago.[114] They came from Alaska where it is believed their ancestors had migrated from Siberia about 45,000 years ago and gradually moved across the central and eastern Arctic of Canada.[115] Their culture, labelled the pre-Dorset culture, extending from around 2000 to 1000 B.C., was oriented to the land rather than the sea and centred on the hunting of caribou. The ensuing Dorset culture, from about 1000 B.C. to A.D. 800, displayed a shift to the hunting of sea mammals, for most known sites are coastal. Around A.D. 1000 the Thule culture began in Alaska and spread rapidly eastward until it had displaced or amalgamated the Dorset culture.[116] The Thule Inuit initially occupied turf dwellings on the coast and pursued large bowhead whales.

About 700 years ago, probably due to a reduction in the availability of whales caused by climatic and sea ice changes, the Thule shifted their living patterns to winter habitation in snow houses on the sea ice, where seals could be hunted at breathing holes, and summer occupation in the interior where fish and caribou could be pursued.[117] This use of the sea ice by the Canadian Inuit was observed by British explorers in search of a Northwest Passage, particularly during the first half of the nineteenth century, and by Canadian explorers at the beginning of the present century.[118] For instance, V. Stefansson described his encounter with a camp of over 50 snow houses on the ice of the Beaufort Sea in the Amundsen Gulf region during his second Arctic expedition (1908–12).[119]

Present-day Inuit do not camp on the sea ice as often and for as long as their ancestors did. However, because of their use of motorized toboggans and skidoos, Inuit hunters now range for much greater distances offshore. A survey of ringed seal hunting areas for the years 1959 to 1974 indicates that 'average areas of around 2000 square miles are not unusual for hunters to cover in the course of sea-ice hunting, with some individuals having hunted over areas in excess of 10,000 square miles',[120] It has been established that all of the 28 Inuit communities across the Canadian Arctic still engage regularly in sea ice hunts in the water areas of the Archipelago including those of Parry Channel. This is particularly so with respect to the western end of Lancaster Sound and the whole of Barrow Strait where Inuit regularly hunt ringed seals and polar bears (see Figure 17).

The foregoing brief review of the vital rights and interests acquired and exercised by the Canadian Inuit, literally since time immemorial, is sufficient to establish that those rights and interests may legitimately be taken into account in support of the validity of the baselines across Lancaster Sound and Amundsen Gulf. In addition, they have been recognized and protected by Canada for such a long period of time that they might well have resulted in a consolidation of title, as was shown to exist for the waters of the Lopphavet basin in the *Fisheries Case*.

11.3. Consolidation of title to Lancaster Sound, Barrow Strait and Amundsen Gulf

An historical consolidation of title goes beyond the existence of certain economic interests evidenced by long usage such as the ones just discussed. As already indicated, such a consolidation involves virtually the same elements as for a case of historic waters except that their proof is not as stringent, since the historical consolidation merely supports or confirms title to a maritime area acquired by the establishment of straight baselines. Those elements are: (1) exercise of State authority; (2) long usage or passage of time; (3) general toleration; and (4) vital interests of Canada as coastal State.

Exercise of State authority

After the transfer of the Arctic islands by Great Britian to Canada, a number of steps were taken not only to consolidate Canada's sovereignty over the land areas of the Archipelago but also to insure a certain control over some of the water areas, particularly on the eastern side of the Archipelago. Most of the measures were directed at the protection of the local population and the resources of the marine environment on which their livelihood depended.

Fig. 17. Polar bear kill locations in Barrow Strait. *Source*: Sikumiut, C.A.R.C., 1984, 91.

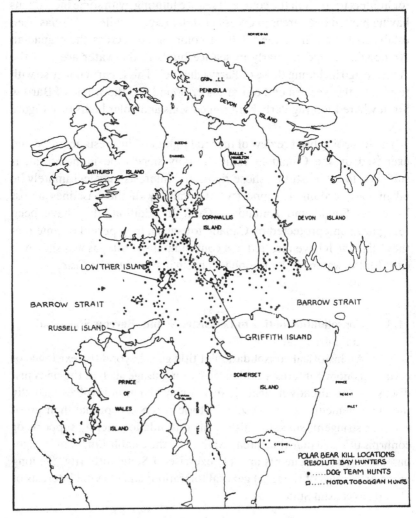

An amendment to the Fisheries Act in 1906, compelling whalers to obtain a licence when hunting whales in Hudson Bay and territorial waters north of the 55th parallel,[121] was enforced in the waters of and adjacent to Lancaster Sound. In 1926, the Arctic Islands Preserve was established by Order in Council which covered all the land and water areas north of the 60th parallel between the 60th and 14th meridians of longitude.[122] The stated purpose of the Preserve was 'to protect both the natives and the wildlife and to place something on the map to indicate that this government control and administer the area'.[123] This was followed by another Preserve and describing it in the same sector form.[124] The enforcement of the game regulations and of the customs laws was done mainly by the Eastern Arctic Patrol which began in 1922, and by the R.C.M.P. Since World War II, the surveillance and patrol of the waters have been assured primarily by the Coast Guard.

In 1970, the Arctic Waters Pollution Prevention Act was adopted specifically to protect the Inuit and their marine environment. The Preamble of the Act states that the Arctic waters are to be navigated only 'in a manner that takes cognizance of Canada's responsibility for the welfare of the Eskimo and other inhabitants of the Canadian Arctic and the preservation of the peculiar ecological balance that now exists in the water, ice and land areas of the Canadian Arctic'.[125] The Act applies to all waters north of the 60th parallel and provides for strict measures to prevent pollution of the marine environment which could occur from the exploitation of resources or shipping. The Act was followed in 1972 by a set of Regulations setting out shipping safety control zones for the whole of the Arctic waters and specifying stringent requirements relating to construction, manning and equipment for ships venturing in these waters (see Figure 18 and Table 2).

In addition to the above legislative and administrative manifestations of State authority, the Territorial Court of the Northwest Territories was established in 1955 and that Court has exercised jurisdiction on several occasions over offences committed on the sea ice of the Archipelago regardless of the distances from the nearest land or island. It did so on the basis that the Northwest Territories Act, as amended in 1952, defined 'Territories' as comprising *inter alia*, 'all that part of Canada north of the Sixtieth Parallel of North latitude'.[126] In 1956, Sissons J. tried an Eskimo charged with murder on the sea ice some 60 miles offshore in Queen Maud Gulf.[127] In 1969, when Canada still had a three-mile territorial sea, Morrow J. tried an Eskimo for unlawfully hunting a female bear with young on sea ice, some ten miles offshore in Larsen Sound on the west side of Boothia Peninsula. The question of jurisdiction having been

Table 2. List of classes for shipping safety control zones.

Category	Zone 1	Zone 2	Zone 3	Zone 4	Zone 5	Zone 6	Zone 7	Zone 8	Zone 9	Zone 10	Zone 11	Zone 12	Zone 13	Zone 14	Zone 15	Zone 16
Arctic Class 10	All Year	All Year	All Year	All Year	All Year	All Year	All Year	All Year	All Year	All Year	All Year	All Year	All Year	All Year	All Year	All Year
Arctic Class 8	Jul. 1 to Oct. 15	All Year	All Year	All Year	All Year	All Year	All Year	All Year	All Year	All Year	All Year	All Year	All Year	All Year	All Year	All Year
Arctic Class 7	Aug. 1 to Sept. 30	Aug. 1 to Nov. 30	Jul. 1 to Dec. 31	Jul. 1 to Dec. 15	Jul. 1 to Dec. 15	All Year	All Year	All Year	All Year	All Year	All Year	All Year	All Year	All Year	All Year	All Year
Arctic Class 6	Aug. 15 to Sept. 15	Aug. 1 to Oct. 31	Jul. 15 to Nov. 30	Jul. 15 to Nov. 30	Aug. 1 to Oct. 15	Jul. 15 to Feb. 28	Jul. 1 to Mar. 31	Jul. 1 to Mar. 31	All Year	All Year	Jul. 1 to Mar. 31	All Year	All Year	All Year	All Year	All Year
Arctic Class 4	Aug. 15 to Sept. 15	Aug. 15 to Oct. 15	Jul. 15 to Oct. 31	Jul.15 to Nov. 15	Aug. 15 to Sept. 30	Jul. 20 to Dec. 31	Jul. 15 to Jan. 15	Jul. 15 to Jan. 15	Jul. 10 to Mar. 31	Jul. 10 to Feb. 28	Jul. 5 to Jan. 15	June 1 to Jan. 31	June 1 to Feb. 15	June 15 to Feb. 15	June 15 to Mar. 15	June 1 to Feb. 15
Arctic Class 3	Aug. 20 to Sept. 15	Aug. 20 to Sept. 30	Jul. 25 to Oct. 15	Jul. 20 to Nov. 5	Aug. 20 to Sept. 25	Aug. 1 to Nov. 20	Jul. 20 to Dec. 15	Jul. 20 to Dec. 31	Jul. 20 to Jan. 20	Jul. 15 to Jan. 25	Jul. 5 to Dec. 15	June 10 to Dec. 31	June 10 to Dec. 31	June 20 to Jan. 10	June 20 to Jan. 31	June 5 to Jan. 10
Arctic Class 2	—	—	Aug. 15 to Sept. 30	Aug. 1 to Oct. 31	—	Aug. 15 to Nov. 20	Aug. 1 to Nov. 20	Aug. 1 to Nov. 30	Aug. 1 to Dec. 20	Jul. 25 to Dec. 20	Jul. 10 to Nov. 20	June 15 to Dec. 5	June 25 to Nov. 15	June 25 to Dec. 10	June 25 to Dec. 20	June 10 to Dec. 10

170

	1	2	3	4	5	6	7	8	9	10	11	12	13
Arctic Class 1A	June 20 to Nov. 30	Jul. 1 to Dec. 10	Jul. 1 to Nov. 30	Jul. 15 to Oct. 31	Jul. 1 to Nov. 10	Jul. 15 to Nov. 10	Aug. 1 to Dec. 10	Aug. 10 to Dec. 10	Aug. 10 to Nov. 20	Aug. 10 to Nov. 5	Aug. 25 to Oct. 31	Aug. 20 to Sept. 15	Aug. 20 to Sept. 30
Arctic Class 1	June 20 to Nov. 15	Jul. 1 to Nov. 30	Jul. 1 to Nov. 30	Jul. 15 to Oct. 15	Jul. 1 to Oct. 31	Jul. 15 to Oct. 20	Aug. 1 to Oct. 31	Aug. 10 to Oct. 31	Aug. 10 to Oct. 31	Aug. 10 to Oct. 15	Aug. 25 to Sept. 30	—	—
Type A	June 20 to Nov. 20	June 25 to Dec. 5	June 25 to Nov. 30	June 25 to Oct. 22	June 15 to Nov. 10	Jul. 10 to Oct. 31	Jul. 25 to Nov. 20	Aug. 1 to Nov. 20	Aug. 1 to Nov. 10	Aug. 1 to Oct. 25	Aug. 15 to Oct. 15	Aug. 20 to Sept. 10	Aug. 20 to Sept. 20
Type B	June 20 to Nov. 10	Jul. 1 to Nov. 30	Jul. 1 to Nov. 30	Jul. 15 to Oct. 15	Jul. 1 to Oct. 25	Jul. 15 to Oct. 20	Aug. 1 to Oct. 31	Aug. 10 to Oct. 31	Aug. 10 to Oct. 31	Aug. 10 to Oct. 15	Aug. 25 to Sept. 30	Aug. 20 to Sept. 5	Aug. 20 to Sept 15
Type C	June 25 to Nov. 10	Jul. 1 to Nov. 25	Jul. 1 to Nov. 25	Jul. 15 to Oct. 10	Jul. 1 to Oct. 25	Jul. 15 to Oct. 15	Aug. 1 to Oct. 25	Aug. 10 to Oct. 25	Aug. 10 to Oct. 25	Aug. 10 to Oct. 10	Aug. 25 to Sept. 25	—	—
Type D	Jul. 1 to Oct. 31	Jul. 5 to Nov. 10	Jul. 10 to Nov. 10	Jul. 30 to Sept. 30	Jul. 1 to oct. 20	Jul. 15 to Oct. 10	Aug. 5 to Oct. 20	Aug. 15 to Oct. 20	Aug. 15 to Oct. 20	Aug. 10 to Oct. 5	—	—	—
Type E	Jul. 1 to Oct. 31	Jul. 20 to Nov. 5	Jul. 20 to Oct. 31	Aug. 15 to Sept. 20	Jul. 1 to Oct. 20	Jul. 15 to Sept. 30	Aug. 10 to Oct. 20	Aug. 20 to Oct. 15	Aug. 20 to Oct. 20	Aug. 10 to Sept. 30	—	—	—

Source: Consolidated Regulations of Canada, 1978, Vol. III, 2271.

Fig. 18. Shipping safety control zones. Prepared for the author by the Canadian Hydrographic Service, Dept of Fisheries and Oceans, Ottawa. *Source:* S.O.R./72–303, *Canada Gazette*, 23 Aug. 1972, Part 11, Vol. 106, No. 16, 1468.

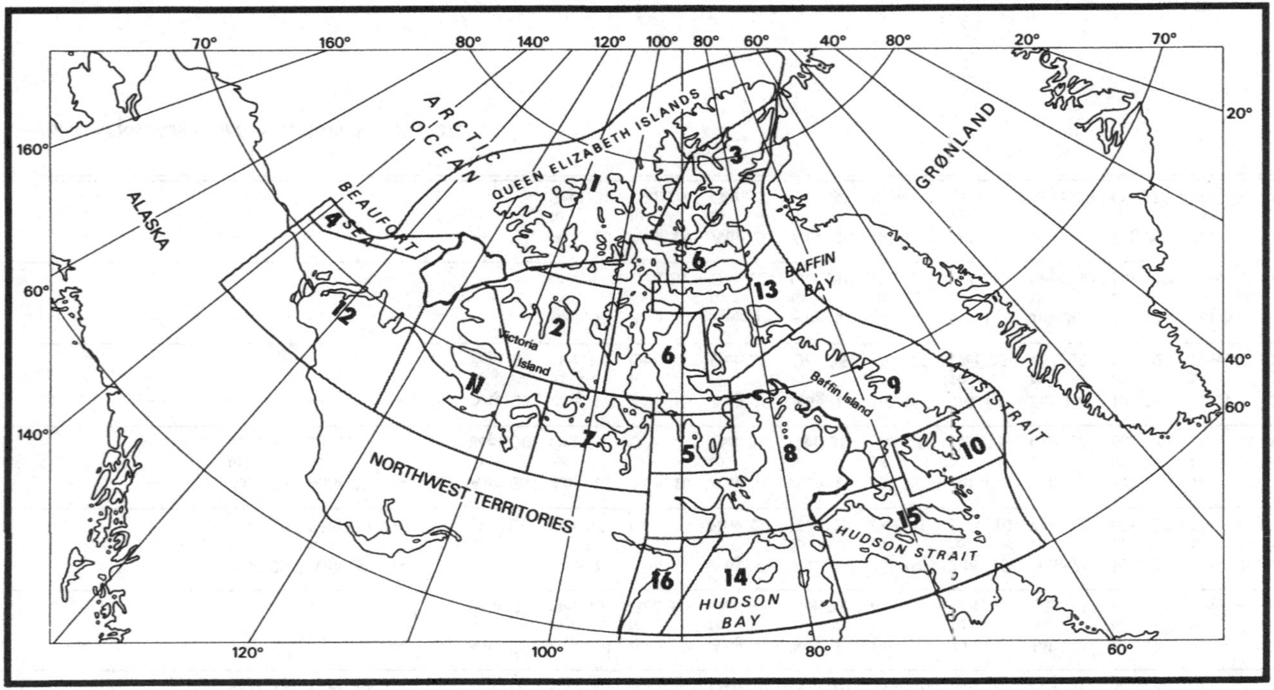

172

raised, Morrow J. traced the history of the definition of 'territories' in the Northwest Territories Act. Noting that, in the first Act of 1906 'Territories' was defined as meaning 'Territories formerly known as Rupert's Land and the Northwest Territory',[128] he held that the new definition of 1952 'in no wise restricts "Territories" to land only as distinct from "land" in the larger sense'[129] and, therefore, his court had jurisdiction. He found in particular that 'the sea-ice, extending off from the land, is within the jurisdiction of the government of Canada.'[130] Addressing himself to the point of jurisdiction over the sea ice, Morrow J. stated:

> If it should happen that the recognition of this jurisdiction over the sea ice has only come in recent times as a result of the comity of nations, or as the result of the activities of the government of Canada in its exercising of sovereignty in these areas, then it still remains that the parliament of Canada in 1952 had to have intended to include the whole area in its definition of 'Territories'.[131]

In 1981, following the same reasoning and quoting extensively from the above judgment, Tallis J. of the Federal Court of Canada held that the term 'Territories' in the Northwest Territories Act permitted the writ of the sovereign to run beyond the low-water mark so as to bring within its ambit of jurisdiction an oil company drilling offshore in the Beaufort Sea.[132]

It follows from the foregoing that Canada has manifested its authority and control over the waters of the Arctic Archipelago through the exercise of legislative, administrative and judicial jurisdiction, and has done so for a long period of time.

Long usage or passage of time
There is no requirement in international law for the passage of a fixed period of time in the appraisal of an historic title properly so-called and, *a fortiori*, there is no such requirement for a mere historical consolidation of title. In the *Fisheries Case*, the consolidation of title to the waters of the Lopphavet took place over a period of over 200 years, but the consolidation of the straight baseline system itself materialized in about 80 years.

In the case of the waters of the Canadian Archipelago, Canada's manifestation of State authority could only begin after it obtained title to the islands in 1880. And, even then, it first had to consolidate its title over some of the northernmost islands and it did so completely by 1930, the year it secured Norway's recognition of sovereignty over the Sverdrup

Islands. Nevertheless, it began exercising legislative and administrative jurisdiction over certain water areas of the Archipelago at the start of this century. Specifically, it enforced whale licencing legislation in Lancaster Sound and adjacent areas beginning in 1906, so that the long usage for those waters is about 80 years. As for the remaining waters of the Archipelago, the various manifestations of State authority do not go back as far and are not as easily related to specific bodies of water, although they are relevant to indicate an intention on the part of Canada to exercise control of waters of the Archipelago generally.

General toleration of States

What is required here is not an acquiescence, as would be the case for historic waters, but simply an absence of opposition on the part of the generality of States[133] or, in the words of the Court in the *Fisheries Case*, a 'general toleration'.[134] With one exception in 1970, which shall now be discussed, there appears to be no recorded case of protest to the various manifestations of State authority by Canada over the waters of the Archipelago. On the contrary, there is evidence of compliance with Canada's laws by foreign whalers, at least in the Lancaster Sound region where Captain J. E. Bernier sold licences to Scottish and American whalers during his voyages of 1906–7 and 1908–9.[135]

The exception of 1970 is the American protest against Canada's extension of territorial waters to 12 miles and the Arctic waters pollution prevention legislation extending 100 miles from the coast. As was explained earlier, this protest was damaging to Canada's subsequent claim of historic internal waters because of the admissions which the protest engendered at the time.[136] The same protest would have little if any effect on a claim of consolidation of title to the waters of Lancaster Sound and Amundsen Gulf in support of the validity of straight baselines across those areas for two reasons. First, a 12-mile territorial sea is now unquestionably permitted by customary international law and, indeed, had become so by 1970 when over 50 States had already adhered to that practice. Second, the objection to the pollution prevention legislation centred on the premise that it constituted an encroachment on the traditional principle of exclusive flag State jurisdiction on the high seas. Assuming for the moment that the waters of those maritime areas were high seas at the time, they would now become part of the exclusive economic zone of Canada; furthermore, the validity of the legislation has been confirmed internationally by the 'ice-covered areas' provision of the 1982 Law of the Sea Convention,[137] which may now be considered as part of customary international law.[138] In short, the 1970 protest cannot be considered an effective protest to prevent a consolidation of title from arising with respect to the areas in question.

Not only has there been a general toleration of Canada's manifestations of State authority in the waters of the Archipelago, but Canada has been considered to have an obligation to take protective measures for the waters of Lancaster Sound, recognized as one of the richest biological areas in the entire Arctic. The Convention for the Protection of the World Cultural and Natural Heritage, to which Canada become a party in 1976,[139] obligates its members to 'take appropriate legal, scientific, technical, administrative and financial measures'[140] to protect 'natural areas of outstanding universal value from the point of view of science, conservation or natural beauty'.[141] And, in 1980, the World Conservation Strategy, adopted by the International Union for Conservation of Nature, designated the Arctic waters as constituting a priority because 'the Arctic environment takes so long to recover'.[142] In addition, the World Heritage Program of the United Nations has recommended that all of Lancaster Sound become a World Heritage Area[143] and Canada is presently studying the implementation of that recommendation. To summarize, suffice it to say for present purposes that there is an international recognition that Lancaster Sound constitutes a unique body of water and that it is up to Canada to decide what precise steps should be taken to insure the protection of the Sound because of the vital interests involved.

Vital interests of Canada as coastal State
Unlike in the case of an historic title, the vital interests of the coastal State form an integral part of the considerations to be taken into account for a consolidation of title. In respect of the Arctic waters, Canada can invoke three types of vital interests, particularly with reference to Lancaster Sound. These relate to the marine environment, the Inuit and to national security.

The marine environment
The waters of Lancaster Sound and vicinity constitute one of the most biologically productive regions of the whole of the Arctic and, at the same time, one of the most ecologically sensitive. The Lancaster Sound region provides a habitat for polar bears, whales (particularly the narwhal and the beluga), seals (harp, ringed, bearded and hooded), walrus and tens of thousands of seabirds of various species. These populations are directly or indirectly dependent upon the productivity of fish, particularly the arctic cod. This is a small fish of no great commercial value but it constitutes the main food of marine mammals and seabirds. The Arctic

cod 'requires 5 years to mature and probably spawns only once in its lifetime, which averages 7 years'.[144] In turn, the fish depends on the continued productivity of microscopic algae called 'phytoplankton', a form of plant life in the oceans on which all other marine-related populations ultimately depend.

The phytoplankton, which drifts on the surface layers, begins to bloom when leads open up in the ice and let the sun and light penetrate. It then continues to reproduce during the short season of open water. If the phytoplankton is subject to an oil spill, the whole food chain is affected. Because of the above, spring and summer are the worst seasons for an oil spill to occur. Professor A. Nelson-Smith, a marine biologist, puts it as follows:

> . . . in spring it [oil spill] might affect the bloom of spontic algae or phytoplankton, with repercussions right up the food-web, or the birds arriving on migration. Seals are, at this time, in a state of physiological stress, with low food reserves, and polar bears are most at risk. An oil spill offshore is most likely to reach the coast quickly during the summer, when inshore waters are ice-free, placing at risk birds which are nesting, foraging or moulting there; later in this season, they will begin to congregate in migration staging-posts.[145]

The special danger for birds is probably greater in the Lancaster Sound region than anywhere else in the Arctic, with the whole of Bylot Island at the southeast entrance of Lancaster Sound being a bird sanctuary. In these circumstances, it is easily understandable that Canada is concerned with the prospects of the Northwest Passage being opened up for year-round navigation by oil tankers. This concern for this vital interest is all the more acute since damage to resources of the marine environment also means damage to its local population.

The Inuit

It has been shown that all 28 Inuit communities of the Canadian Arctic depend on hunting and fishing for their livelihood and subsistence, but this is particularly so for the old community of Pond Inlet on the south side of Lancaster Sound and the settlement of Grise Fiord on the north side. The hunters of Grise Fiord covered the largest ringed seal hunting area during the years of 1959 to 1974.[146] As was demonstrated earlier,[147] not only does the harvesting of marine-related resources constitute a vital need for the survival and welfare of the Inuit, but the continuation of their whole way of life and culture are dependent upon those resources.

Canada has a strong moral obligation to protect its Inuit population and, indeed, it might even have a legal obligation. As a party to the International Covenant on Civil and Political Rights since 1976, Canada has obligated itself not to deny its ethnic minority groups 'the right, in community with the other members of their group, to enjoy their *own culture*, to profess and practice their own religion, or to use their own language'.[148] Whatever the precise extent of that international obligation, Canada has already taken certain protective measures in this regard[149] and it has a vital interest in insuring that the Inuit continue to enjoy their traditional way of life and culture peacefully.

National security
Should the Northwest Passage be developed and used for international navigation, which appears to be virtually inevitable, the new right of transit passage for both commercial and warships would apply, including the right of submerged passage for submarines.[150] When one considers the strategic position of the Canadian Arctic, located between the two super powers, and the feasibility of under-ice navigation, one can understand the potential threat to Canada's security.[151] Both the United States and the Soviet Union have fully tested the Arctic waters with their respective submarines and the Soviet Union's Northern Fleet submarines are based on the Kola Peninsula.

The waters of the Canadian Arctic Archipelago may be entered by submarine from a number of directions, in particular from Robeson and Kennedy Channels between Greenland and Ellesmere Island, then into Lancaster Sound. A submarine may also enter Parry Channel by M'Clure Strait, and there is enough depth throughout Parry Channel for a submarine to go through without difficulty. This was established by the westerly crossing of Parry Channel in 1960 by the U.S.S. *Seadragon*. In these circumstances, Canada has a vital interest in exercising the fullest surveillance and control over Lancaster Sound, the eastern gateway of the Northwest Passage. It can only do so if those waters are internal waters.

11.4.   Summary
1. The Canadian Arctic Archipelago is a huge labyrinth of islands and headlands, broken by large indentations in the form of bays and gulfs, traversed by numerous straits, sounds and inlets, and the whole being fused together by an ice cover most of the year.
2. Although the Archipelago is not a simple 'fringe of islands' in the

strict sense, it constitutes a single unit bordering the northern coast of Canada and forming an integral part of the coast.

3. The physical characteristics of the coast and of the Archipelago are such as to make it absolutely impossible to follow the sinuosities of the coast or of the islands in the measurement of the territorial sea and render it necessary to use straight baselines.

4. The newly established straight baselines meet the two flexible criteria or guidelines formulated in the *Fisheries Case* and incorporated as compulsory criteria in the 1958 and 1982 Conventions: the general direction of the coast and the close link between the sea and the land areas.

   (1) 'general direction': what really constitutes the Canadian coastline being the outer line of the Archipelago, the straight baselines follow such outer line.

   (2) 'close link': the sea to land (islands) ratio in the Archipelago is 0.822 to 1 and this close relation is enhanced by the quasi-permanent presence of the ice cover in the enclosed water areas.

5. The practice of States in general and that of four of the five Arctic States in particular reflect a liberal interpretation of the geographic requirements for the applicability of straight baselines and lend strong support to the baselines drawn around the Canadian Arctic Archipelago.

6. The validity of the straight baselines crossing Amundsen Gulf and Lancaster Sound is reinforced by taking into account the economic interests of the local Inuit population whose livelihood has depended exclusively on fishing, hunting and trapping in those water areas since time immemorial.

7. Although there is no limit to the length of individual straight baselines established in conformity with the compulsory guidelines, particularly when their validity is reinforced by regional economic interests evidenced by long usage, an effort has been made to restrict the length of the baselines so that the average length is less than 17 miles and the longest is 99.5 miles.

8. As an additional basis to the straight baseline system for claiming the waters of the Archipelago as internal waters, Canada is in a position to rely on an historical consolidation of title to Lancaster Sound and Barrow Strait in the east and Amundsen Gulf in the west, the effect of such consolidation being to confirm or consolidate the title to those water areas resulting from their enclosure by straight baselines.

9. The consolidation of title to Lancaster Sound, Barrow Strait and Amundsen Gulf arises out of the exercise of legislative and administrative jurisdiction by Canada for about 80 years and the general toleration of States, the purpose of such exercise being to protect certain vital interests in those water areas.

10. The vital interests of Canada in the waters of the Archipelago in general, and those of Lancaster Sound, Barrow Strait and Amundsen Gulf in particular, reside in its obligation to protect three national concerns: the unique marine environment, the traditional livelihood and culture of the Inuit and the security of its territory.

Notes to Part 3

1. See Art. 4, Convention on the Territorial Sea and the Contiguous Zone, A/CONF. 13/L.52 (1958) and Art. 7, Convention of the Law of the Sea, A/CONF. 62/122 (1982). For greater convenience, those conventions will now be referred to simply as the 1958 Convention and the 1982 Convention.
2. See A/CONF. 62/L.4 (26 July 1974).
3. A/CONF. 62/C.2/L.51 (12 Aug. 1974).
4. Art. 131, *I.S.N.T.*, A/CONF. 62/WP.8 (7 May 1975).
5. Preamble, *supra* note 1. A similar provison appeared in the 1958 Convention.
6. [1951] I.C.J. Rep., at 128–9.
7. See Art. 4, para. 1, 1958 Convention and Art. 7, para. 1 of the 1982 Convention.
8. For a description of the skjaergaard, see *supra* note 6, at 127.
9. See O'Connell, *The International Law of the Sea*, Vol. I, at 212, where he gives some 18 such examples.
10. See, in particular, the *North Sea Continental Shelf Cases* (1969) I.C.J. Rep., at 31, where adjacency was held to imply only proximity in a general sense.
11. *Supra* note 6, at 129; emphasis added.
12. *Ibid.* at 131; emphasis added.
13. Art. 3, 1958 Convention.
14. See Articles 5 and 7 of the 1982 Convention.
15. *Supra* note 9, at 209; emphasis added.
16. Charles De Visscher, *Theory and Reality in Public International Law* (1957), at 217; emphasis added. The present writer had previously viewed the straight baseline system as an exception (see *The Law of the Sea of the Arctic, with Special Reference to Canada*, 1973 at 72); however, upon further analysis of the Court's judgment and Judge De Visscher's writings, he has now come to a different conclusion.
17. *Supra* note 6, at 133; emphasis added.
18. *Ibid.*, emphasis added.
19. See Art. 4, para. 2, 1958 Convention and Art. 7 para. 3, 1982 Convention.
20. *Supra* note 6, at 129. This was published as an *erratum* on October 22, 1956, and should be inserted in the (1951) I.C.J. Rep., English text, at 129.
21. *Ibid.*, at 142.
22. *Ibid.*
23. *Ibid.*, emphasis already in text.
24. *Ibid.*, at 133; emphasis added.
25. *Ibid.*

26. Art. 5, para. 2, 1958 Convention and Art. 8, para. 2, 1982 Convention.
27. *Supra* note 6, at 133.
28. At. 4, para. 4, 1958 Convention and Art. 7, para. 5, 1982 Convention.
29. *Supra* note 6, at 142.
30. *Ibid.*
31. *Ibid.*, at 133.
32. *Ibid.*, at 142.
33. *Ibid.*
34. *Ibid.*, at 131.
35. *Ibid.*
36. *Ibid.*
37. *Ibid.*
38. *Supra* note 1, Art. 47, para. 2.
39. 4 A.J.I.L. 226–36 (1910). See also the original French text of the decision in United Nations, *Reports of International Arbitral Awards*, Vol. IV, 155–62.
40. *Ibid.*, at 230.
41. *Ibid.*, at 231–2.
42. *Ibid.*, at 232.
43. *Ibid.*, at 233.
44. *Ibid.*
45. *Ibid.*, at 233–4.
46. *Supra* note 16, at 200; emphasis added.
47. *Ibid.*, footnote 69. This principle was also endorsed by Judge Ammoun in the *North Sea Continental Shelf Cases* (1969) I.C.J. Rep., at 113.
48. Charles De Visscher, *Problèmes d'interprétation judiciaire en Droit international public*, at 176, footnote 2 (1963).
49. *Supra* note 6, at 138.
50. *Ibid.*, at 139; emphasis added.
51. *Supra* note 16, at 200. For a discussion of the scope of application of this doctrine, see D. H. N. Johnson, 'Consolidation as a Root of Title in International Law', 13 *Camb. L. J.* 215–25. (1955).
52. *Supra* note 39, at 234.
53. *Supra* note 16, at 200; emphasis added.
54. *Ibid.*
55. De Visscher, *supra* note 16, at 201; emphasis added.
56. *Supra* note 39, at 234.
57. *Supra* note 6, at 138; emphasis added.
58. *Ibid.*
59. *Ibid.*
60. *Supra* note 39, at 234.
61. Reproduced in the decision of the Court, *supra* note 6, at 125; emphasis added.
62. *Ibid.*, at 136; emphasis added.
63. *Ibid.*, at 142.
64. *Ibid.*, at 139.
65. P. B. Beazley, 'Territorial Sea Baselines', *International Hydrographic Review*, Vol. 48, at 145 (1971).
66. The writer is indebted to John Cooper, formerly Territorial Waters Officer with the Canadian Government, who prepared this Table.
67. *Supra* note 6, at 134.
68. *Ibid.*
69. *Ibid.*, at 143.
70. *Ibid.*, at 142.

71. Decree dated 30 June 1955, *in* F. Durante and W. Rodino (eds), *Western Europe and the Development of the Law of the Sea*, Binder II, at 39 (1980).
72. *Ibid.*, at 123.
73. Decree dated 17 December 1976, *ibid.*, at 251.
74. Art. 2, Order No. 191, dated 27 May 1963, *supra* note 71, at 31.
75. Arts. 3 and 4, Order No. 437, dated 21 December 1966, *supra* note 71, at 73.
76. Art. 1, Order No. 629, dated 22 Dec. 1966, *supra* note 71, at 155.
77. U.S. Dep't of State, 'Sovereignty of the Sea', *Geographic Bulletin*, No. 3, at 12 (April 1965).
78. *Ibid.*, at 13.
79. See Whiteman, *Digest of International Law*, Vol. 4, at 287 (1965).
80. *Ibid.*, at 283 and 284.
81. Arthur H. Dean. 'The Second Geneva Conference on the Law of the Sea: The Fight for Freedom of the Seas', 54 *A.J.I.L.* 751, at 753 (1960).
82. *Supra* note 80, at 281.
83. See Articles 47 and 53 of the 1982 Convention.
84. Art. 46, *ibid.*
85. *Digest of United States Practice in International Law, 1978*, at 943 (1980); emphasis added.
86. Art. 5, Law on the State Boundary of the U.S.S.R., entered into force March 1, 1983; see English translation in 22 *Int'l Legal Materials* 1055–76, at 1057 (1983).
87. See amendment of June 10, 1971 to the 1960 Statute on the Protection of the State Border of the U.S.S.R., in Wm. E. Butler, 'New Soviet Legislation on Straight Baselines', 20 *I.C.L.Q.*, at 751 (1971).
88. See Declaration No. 4450, reproduced in 21 *Notices to Mariners* (in Russian), at 27 (1985).
89. *Statutes of Canada*, 1964, c. 22, s. 5.
90. See Orders in Council, P.C. 1967, at 2025 (26 Oct. 1967) and P.C. 1969, at 1109 (26 May 1969). Although these were Canada's first straight baselines properly so-called, a baseline was drawn across the entrance of Hudson Strait in 1937; see P.C. 3139, dated 18 December 1937.
91. See the following: Shipping Safety Control Zones Order, S.O.R./72–303, *Canada Gazette* Part II, Vol. 106, No. 16, at 1468 (23 Aug. 1972); Agreement between Canada and Denmark, *Canada Treaty Series* 1974, No. 9; and S.O.R./77–173, *Canada Gazette* Part II, Vol. 111, No. 5, at 652 (9 March 1977).
92. See Territorial Sea Geographical Coordinates (Area 7) Order, S.O.R./85–872, *Canada Gazette* Part II, Vol. 119, No. 20, at 3996 (10 Sept. 1985).
93. Statement in the House of Commons by Secretary for External Affairs Joe Clark, 10 Sept. 1985, reproduced in *Statement* Series 85/49, at 3.
94. See Pharand, *The Law of the Sea of the Arctic, with Special Reference to Canada*, at 88–93 (1973) where the following are reviewed: M. Cohen, 'Polar Ice and Arctic Sovereignty', 73 *Saturday Night*, No. 18, at 35 (30 Aug. 1958); Ivan L. Head, 'Canadian Claims to Territorial Sovereignty in the Arctic Regions', 9 *McGill L. J.* 201, at 218 (1962–3); J. -Y. Morin, 'Le progrès technique, la pollution et l'évolution récente du droit de la mer au Canada, particulièrement à l'égard de l'Arctique', 8 *C.Y.I.L.* 158, at 240 (1970); M. Cohen, 'The Arctic and the National Interest', 26 *Int'l J.* 52, at 80 (1970–1); W. G. Reinhard, International Law: Implications of the Opening of the Northwest Passage', 74 *Dickinson L. Rev.* 678, at 688–9 (1970) and J. W. Dellapenna, 'Canadian Claims in Arctic Waters', 7 *Land and Water L. Rev.* 383, at 420 (1972). The writer's own examination of the question appears at 93–98.
95. *Supra* note 6, at 139.
96. *Ibid.*, at 129.

97. *Supra* note 7.
98. *Supra* note 6, at 132.
99. See *supra* note 9, at 212.
100. *Ibid.*, at 209.
101. *Supra* note 6, at 127.
102. For a similar conclusion, see W. G. Reinhard, 'International Law: Implications of the Opening of the Northwest Passage', 74 *Dickinson L. Rev.* 678, at 688 (1970) and J. W. Dellapenna. 'Canadian Claims in Arctic Waters', 7 *Land and Water L. Rev.* 383, at 418 (1972).
103. This ratio represents the water area (366,862 square statute miles) as compared to the land area (445,814 square statute miles), those areas being measured north of the Arctic circle. It is important to note that the calculation of land areas is restricted to islands and does not include the northern coastal strip above the Arctic circle.
104. See a statement to this effect by Prime Minister Trudeau, in *Can. H. C. Debates*, Vol. I, at 39 (1969) and a written assurance in the Canadian Note to the United States in 1970, reproduced in *Can. H. C. Debates*, Appendix, at 6028 (17 April 1970).
105. *Supra* note 6, at 142.
106. M. R. Freeman (ed.), *Report: Inuit Land Use and Occupancy Project*, 3 Vols, Can. Gov. Cat. No. R2–46/1976. On the present use of the sea ice, see M. R. Freeman, *Contemporary Inuit Exploitation of the Sea Ice Environment*, paper presented at the Sikumiut Workshop, McGill University, 15 April 1982.
107. W. D. Brackel, 'Socio–Economic Importance of Marine Wildlife Utilization', *Beaufort Sea Technical Report* No. 32, at 36 (1977).
108. P. Usher, 'Renewable resource development in Northern Canada', in *Northern Transitions*, at 154 (1978).
109. H. Myers, 'Traditional and Modern Sources of Income in the Lancaster Sound Region', 21 *Polar Record* 11, at 13 (No. 130, Jan. 1982). See also *Beaufort Sea Environmental Impact Statement*, Vol. 3B, Table 2.51.
110. *Ibid.*
111. Affidavit of Rosita F. Worl, in Plaintiff's Memorandum in Opposition to Defendants' Motion for Judgment on the Pleadings, *Inupiat Community of the Arctic Slope et al. v. The United States et al.*, Civil Action No. A81–0819, U.S. District Court of Alaska (1983).
112. O. Schaefer and J. Steckle, *Dietary Habits and Nutritional Base of Native Populations of the Northwest Territories and Research* (Aug. 1980).
113. M. R. Freeman, *Contemporary Inuit Exploitation of the Sea Ice Environment*, paper presented at the Sikumiut Workshop. McGill University, 15 April, 1982, at 13; emphasis added.
114. See Peter Schledermann, 'Inuit Prehistory and Archaeology', *in* M. Zaslow (ed.), *A Century of Canada's Arctic Islands* 245–56, at 253 (1981).
115. For a summary of Inuit pre-history, see *Baker Lake v. Minister of Indian and Northern Affairs* (1980) 1 Federal Court Reports 518, at 528–32.
116. *Ibid.* at 529.
117. See C. Arnold, 'Archaeology in the Northwest Territories', *Northern Perspectives*, Vol. 10, No. 6 (Nov.–Dec. 1982).
118. See Alan Cooke, *Historical Evidence for Inuit Use of the Sea Ice*, paper presented at the Sikumiut Workshop, McGill University, 15 April 1982.
119. *Ibid.*, at 5–6.
120. Freeman, *supra* note 114, at 9.
121. 5 Edw. VII, c. 13, 1906 Statutes of Canada.
122. P.C. 1146, dated 19 July 1926, *Canada Gazette*, 31 July 1926, at 382.

123. Memorandum attached to P.C. No. 1146.
124. P.C. 807, *Canada Gazette*, 25 May 1929, at 4018.
125. C. 47, 18–19 Eliz. II, 1969–70.
126. S. 2(i), *Northwest Territories Act*, R.S.C. 1952, c. 33.
127. The case is mentioned as a precedent in *Regina v. Tootalik E4–321* (1969) 71 W.W.R. 435, at 440.
128. Quoted in *ibid.*, at 442.
129. *Ibid.*, at 443.
130. *Ibid.*
131. *Ibid.*
132. *B.P. Exploration Co. (Libya) Ltd. v. Hunt* (1981) 1 W.W.R. 209, at 242–5.
133. See De Visscher, *supra* note 16, at 201.
134. *Supra* note 6, at 138 and 139.
135. See J. E. Bernier, *Master Mariner and Arctic Explorer* 312, 313 and 319.
136. See 'Appraisal of Canada's Claim of Historic Waters', Chapter VIII, *supra.*
137. Art. 234, A/CONF. 62/122 (1982).
138. See Donat Pharand, 'La contribution du Canada au développement du droit international pour la protection du milieu marin: le cas spécial de l'Arctique', 11 *Etudes internationales* 441–466 (1980).
139. *Canada Treaty Series*, 1976, No. 45.
140. *Ibid.*, Art. 5(d).
141. *Ibid.*, Art. 2.
142. International Union for the Conservation of Nature, *World Conservation Strategy: Living Resource Conservation for Sustainable Development*, chapter 19, paragraph 12 (1980).
143. *Beaufort Sea – Mackenzie Delta Environmental Impact Statement*, Vol. 3B, at 2.92 (1982).
144. A. Nelson-Smith, 'Biological Consequences of Oil-Spills in Arctic Waters', *in* Louis Rey (ed.), *The Arctic Ocean* 275–93, at 285 (1982).
145. *Ibid.*, at 286.
146. See Table 1 in Freeman, *Supra* note 114, at 10.
147. See 'Regional Economic Interests Evidenced by Long Usage' in Section 2 of the present Chapter.
148. Art. 27.
149. See in particular the explanatory memo attached to the Arctic Islands Preserve of 1926, P.C. 1146 and the Preamble of the Arctic Waters Pollution Prevention Act of 1970, c. 47, 18–19 Eliz. II, 1969–70.
150. This point is developed in Chapter 14 *infra* on the Legal Status of the Northwest Passage.
151. This aspect is developed by the present writer in *The Northwest Passage: Arctic Straits*, Martinus Nijhoff Publishers, 236 pages, at 144–159 (1984); see also W. Harriet Critchley, 'Polar Deployment of Soviet Submarines', 39 *International Journal* 828–65, at 860–5 (1984).

Part 4
The Waters of the Canadian Arctic Archipelago and the Northwest Passage

Regardless of the precise legal status of the waters of the Canadian Arctic Archipelago generally, a special study of the legal regime applicable specifically to the Northwest Passage is necessary.[1] This is particularly so having regard to the eventual use of the Passage for the transportation of hydrocarbon resources from the Beaufort Sea and the Archipelago itself. This part will describe the Northwest Passage, review its past, present and potential use and, finally, analyse the legal status of the Passage.

12

Description of the Northwest Passage

According to the Sailing Directions of Arctic Canada, 'the Northwest Passage spans the North American Arctic from Davis Strait and Baffin Bay in the east to Bering Strait in the west'.[2] Although this represents the traditional definition of the Northwest Passage, the present discussion will limit the meaning of the Passage to the constricted waters within the Canadian Arctic Archipelago between Baffin Bay in the east and the Beaufort Sea in the west. Under the present definition, the latter bodies of water form part of the approaches to the Northwest Passage.

12.1.   Approaches to the Northwest Passage
Eastern approaches

   The Labrador Sea, Davis Strait and Baffin Bay constitute the eastern approaches to the Northwest Passage.[3] The ice conditions in Davis Strait and Baffin Bay are mainly controlled by a warm current flowing north along the west coast of Greenland and a cold current flowing south along the coast of Baffin Island, as well as by a major polynya or open water area at the northern end of Baffin Bay.

   A special feature affecting the navigation conditions of the eastern approaches is the presence of icebergs, calved mostly from the northwestern part of Greenland but partly also from smaller glaciers on Ellesmere and Devon Islands. Icebergs drift southward mainly with the cold Canadian current and, consequently, are found in greater numbers along the Canadian coast. Some of the glaciers rise to 300 feet above sea level and constitute a considerable hazard to navigation. In these circumstances, the recommended route to enter the eastern end of the Northwest Passage is to follow the Greenland coast up to about 74 degrees North latitude and then cross Baffin Bay to enter Lancaster Sound on its northern side. This is the route which the proponents of the Arctic Pilot

Project originally planned to follow for the transportation of liquefied natural gas from Melville Island.

Freeze-up in North Baffin Bay begins in mid-September and spreads southward along the Baffin Island coast during October, developing into an unmoving ice cover as far as the Labrador Sea during the winter, when the ice averages more than one metre in thickness. The spring break-up in Northern Baffin Bay is greatly facilitated by the 'North water' or polynya in Smith Sound, which expands southward during the month of June, thus opening the entrance to Lancaster Sound, usually by the end of that month.

Western approaches

The approaches to the western end of the Northwest Passage are comprised of Bering Strait, Chukchi Sea and Beaufort Sea. Because they are situated south of the normal reach of the permanent polar ice pack, Bering Strait and Chukchi Sea generally offer more favourable ice conditions than does the Beaufort Sea. There is also a recurring winter lead at the edge of the landfast ice along the northwest coast of Alaska between Point Hope and Point Barrow, presenting open water normally from late May to early November.

As for the ice in the Beaufort Sea itself, the polar pack does not generally reach the continental coast of Canada and Alaska, but that coast does not always offer open water. In winter, there is a belt of fast ice along the coast extending several miles offshore to approximately the 20-metre depth contour and this does not generally break up until the end of July. Between the fast ice and the polar pack ice, there is a transition zone of first-year ice which may vary from 50 to 100 miles wide, forming early in October and breaking up at the end of June. This middle zone generally widens upon approaching Amundsen Gulf where an open water area usually forms off Cape Bathurst during March and April, thus accelerating the disintegration of the first-year ice. There is seldom a clear demarcation between this middle zone of first-year ice and the polar pack to the north and, indeed, a number of polar pack floes, some of them hummocked, are often mixed with the first-year ice. Ice conditions may vary from year to year and sometimes it is possible to navigate between Point Barrow and Amundsen Gulf without seeing any ice from late July to early September. Freeze-up in the Beaufort Sea generally occurs at the beginning of October but may not take place until early November if the polar pack is well offshore.

Aside from the polar pack floes which a ship may encounter, the Beaufort Sea presents a hazardous region off the Tuktoyaktuk Peninsula, where numerous submarine pingo-like features are found. These features are much like the well-known pingos (conical-shaped mounds with a central core of ice) on the Tuktoyaktuk Peninsula and they were probably formed in a similar way. They vary from 15 to 45 metres in height and range from about 500 to 1000 metres in diameter at their base. By 1971, some 178 such features had been located in the Beaufort Sea. In 1983, the Canadian Hydrographic Service cleared a 10-mile wide shipping corridor across the pingo area and located more than 20 such pingos.

12.2. Main routes of the Northwest Passage
The Northwest Passage consists of five basic routes, plus at least two variations of two of those routes (see Figure 19). All of the routes are potentially feasible for navigation, although not necessarily for deep draft ships. The choice of the route will depend mainly on the ice conditions,[4] on the size of the ship, the latter's icebreaking capability or that of the accompanying icebreaker, if any, and the adequacy of hydrographic surveys.

Of the five basic routes, only Routes 1 and 2, for the moment, are known to be suitable for deep draft navigation. Routes 3 (and its variation 3A) and 4, which follow the continental coast of Canada for more than half the distance, are not suitable for deep draft navigation. Route 5 presently leads only to Routes 3 and 4, but if it should prove possible eventually to avoid certain shoal areas in Fury and Hecla Strait, its variation Route 5A could lead to Routes 1 and 2 and be used by deep draft ships.

Route 1, through Prince of Wales Strait
Route 1 is presently the most feasible for icebreaking tankers and has been chosen by Dome Petroleum as its primary route. It lies across five bodies of water: Lancaster Sound, Barrow Strait, Viscount Melville Sound, Prince of Wales Strait and Amundsen Gulf.

Lancaster Sound is a steep-sided strait, with an entrance of 51 miles, and extends westward 160 miles to Prince Leopold Island. The ice does not normally consolidate completely in Lancaster Sound and the most favourable period for navigation is between mid-July and the end of October.

Fig. 19. (*Overleaf*) Main routes of the Northwest Passage. Prepared for the author by the Canadian Hydrographic Service, Dept of Fisheries and Oceans, Ottawa.

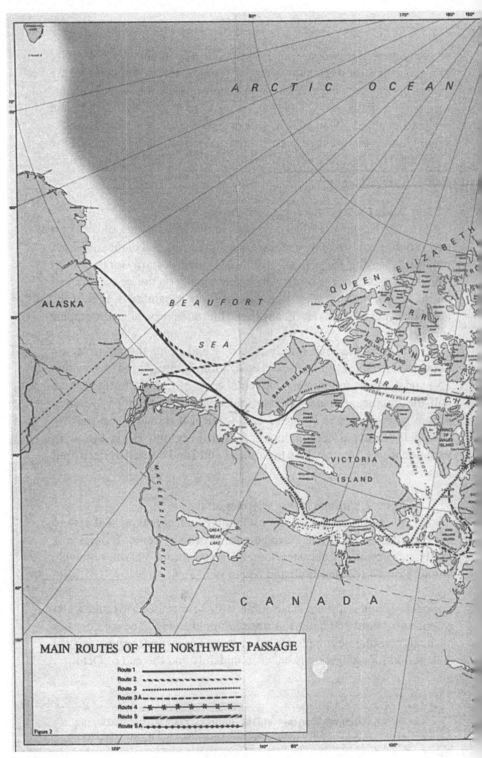

MAIN ROUTES OF THE NORTHWEST PASSAGE

Route 1
Route 2
Route 3
Route 3A
Route 4
Route 5
Route 5A

Figure 2

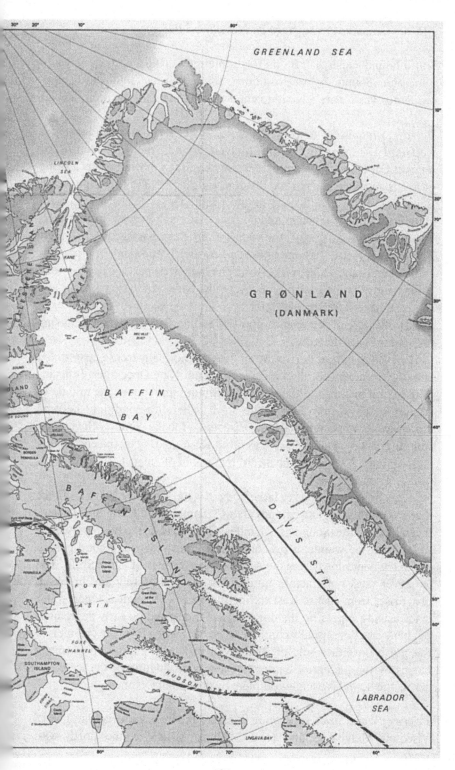

South-bound icebergs in Baffin Bay sometimes make a partial entry into Lancaster Sound before resuming their journey southward; consequently, vessels are advised to enter the Sound on its northern side.

Barrow Strait is 28 miles wide at its eastern entrance and extends westward 170 miles to Viscount Melville Sound, where it widens to 66 miles at Cape Cockburn. At its western end, Barrow Strait offers a string of five islands lying across the Strait in a zigzag fashion (see Figure 20). Beginning on the south side of the Strait, these islands are Russell, Hamilton, Young, Lowther and Garrett. Although Barrow Strait itself is 70 miles wide at this point, the widest inter-island passage is 15.5 miles. The latter is Kettle Passage between Lowther Island lying in mid-strait and Young Island to the south-west. The narrowest passage lies between Young and Hamilton Islands and is 8 miles wide. All these passages and channels have sufficient depth for deep draft ships; however, there exist shoals off most of the islands and the Sailing Directions advise giving these islands a wide berth.[5]

The eastern part of Barrow Strait does not normally freeze up until the end of November, nor does it consolidate until late December. It begins to break up in July and the process continues until about the middle of August in the western half of the Strait, where ice conditions become more severe when approaching Viscount Melville Sound. Navigation from the east is usually feasible as far as Resolute Bay by the end of July and, for the full length of the Strait, by the end of August.

Viscount Melville Sound is a large body of water with an eastern entrance that is 60 miles wide and extends about 220 miles to M'Clure Strait where the width of the Sound is 74 miles. The solid ice cover, prevailing throughout the winter, is predominantly multi-year ice, a good part of which has come down from the Arctic Ocean through M'Clure Strait. The ice cover does not fracture until about the end of July and break-up progresses from east to west during the month of August. Open water leads usually appear on the north side of the Sound by early September but heavy ice floes usually continue to be discharged from the channels to the north on either side of Byam Martin Island. Freeze-up normally begins by the middle of September and consolidation of the ice cover across the Sound is usually complete by the end of October.

Prince of Wales Strait, separating Banks Island from Victoria island, links Viscount Melville Sound (and M'Clure Strait) to the north with

Fig. 20. West Barrow Strait and Prince of Wales Strait. Prepared for the author by the Canadian Hydrographic Service, Dept of Fisheries and Oceans, Ottawa.

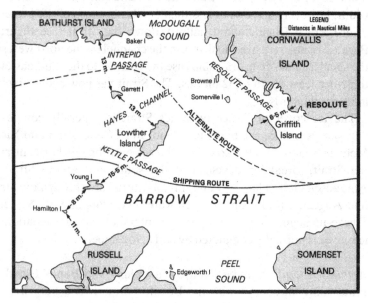

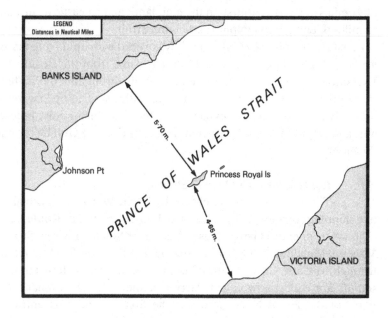

Amundsen Gulf to the south. The Strait is some 145 miles long, averages about 10 miles wide and has a minimum width of 6.5 miles. This minimum width, however, narrows to less than 6 miles because of the presence of Prince Royal Islands about halfway down the Strait (see Figure 20). Because of an extensive shoal area in the channel on the northwest side of the small islands, deep draft ships use the channel to the southeast, where there is a least depth of 32 metres. The Strait has now been surveyed to modern standards.

The ice cover of Prince of Wales Strait is generally composed of first-year ice, since the prevailing winds in M'Clure Strait and Viscount Melville Sound tend to carry the polar ice past the northern entrance of the Strait. Break-up develops at the south end of the Strait from Amundsen Gulf in late July and spreads northward through the Strait by mid-August, with maximum open water occurring in early September. Freeze-up begins in the latter part of September and consolidation of the ice cover is normally completed by early November.

Amundsen Gulf, lying between the mainland and Banks Island, links Prince of Wales Strait with the Beaufort Sea. It is a very irregularly shaped gulf, approximately 200 miles in length, with a western entrance of 91 miles at Cape Bathurst on the mainland, and an eastern entrance of 55 miles opening into Dolphin and Union Strait.

In winter, Amundsen Gulf is usually covered with unmoving first-year ice, with the odd polar ice floe which has come from the Beaufort Sea. Exceptionally, the ice remains virtually mobile throughout the winter due to the presence of a polynya around Cape Bathurst. This open water area normally forms in May and expands into the Gulf during July. Freeze-up normally begins in early October and spreads over the Gulf by the end of the month.

Route 2, through M'Clure Strait
This route, which was chosen by Dome Petroleum as an alternative shipping corridor, follows the same bodies of water as Route 1 except at its western end. Having crossed Lancaster Sound, Barrow Strait and Viscount Melville Sound, Route 2 reaches the Beaufort Sea by continuing through M'Clure Strait north of Banks Island. Those four straits and sounds are collectively called Parry Channel, after Lieutenant W. E. Parry of the Royal Navy, who was the first to reach the entrance of M'Clure Strait in 1819. As for M'Clure Strait itself, it was named after Commander Robert M'Clure, also of the Royal Navy, who became the first to cross (albeit on foot) the unsailed part of the Strait in 1853.

The eastern entrance of M'Clure Strait is 74 miles wide at Cape Providence on Melville Island, and extends approximately 165 miles to the Beaufort Sea where the width of the entrance is nearly 100 miles.

Until July, the strait is normally covered by consolidated ice, most of which is multi-year ice from the Arctic Ocean. Break-up normally occurs during August but the strait usually remains choked with polar ice throughout most of the summer. Freeze-up usually begins around the middle of September and the ice consolidates by the end of October or early November. This is the strait which the S/T *Manhattan* attempted to navigate in 1969, but without success.

Route 3, through Peel Sound and Victoria Strait
Proceeding from east to west, this third route crosses the following eleven bodies of water: Lancaster Sound, Barrow Strait, Peel Sound, Franklin Strait, Larsen Sound, Victoria Strait, Queen Maud Gulf, Dease Strait, Coronation Gulf, Dolphin and Union Strait and Amundsen Gulf. Lancaster Sound, Barrow Strait and Amundsen Gulf having already been described as part of Route 1, the present description will be confined to the remaining eight bodies of water.

Peel Sound is a strait which is approximately 15 miles wide, separates Prince of Wales Island from Somerset Island, and runs from Barrow Strait to Bellot Strait where it joins Franklin Strait to the south. Peel Sound is covered with a solid sheet of ice in winter, this being mainly first-year ice, but with a small inflow of older ice originating in Barrow Strait. Break-up normally begins during the first week of August and is completed by the first week of September. Freeze-up usually begins at the end of September and consolidation of the ice cover occurs during the month of October.

Franklin Strait, separating the southeast coast of Prince of Wales Island from the northern part of Boothia Peninsula, runs from Bellot Strait to Larsen Sound and is about 20 miles wide. The ice conditions are virtually identical to those of Peel Sound.

Larsen Sound is a large body of water lying between Victoria Island and Boothia Peninsula. It leads into James Ross Strait, on the east side of King William Island, and M'Clintock Channel on the west side of that island. Break-up normally begins during the last week of July and the Sound should be clear of ice by the end of August. Freeze-up usually begins at the end of September, with a solid ice cover forming at the end of October.

Victoria Strait links Larson Sound and M'Clintock Channel to the north, with Queen Maud Gulf to the south. It is a wide strait, with the relatively large Royal Geographical Society Islands at its southern extremity upon entering Queen Maud Gulf. Adequate survey has not yet been completed, but charted soundings already indicate several patches of 10-metre depth only. The Strait is covered with rough ice most of the year, of which some is polar ice that has drifted down from Viscount Melville Sound through M'Clintock Channel. Significant break-up does not begin until the end of July and freeze-up develops during the latter part of September.

Queen Maud Gulf, which has an eastern entrance only eight miles wide, quickly widens into a large body of water extending some 170 miles to the west along the continental coast as far as Dease Strait, where the entrance is also only eight miles wide. Navigation is made rather difficult in a number of places in the Gulf because of the presence of numerous islands, reefs and shoals. The ice conditions in Queen Maud Gulf are described in the Sailing Directions as the most difficult on this route, east of Cape Parry on the south shore of Amundsen Gulf.[6] Break-up begins during the last week of July but does not end until the middle of August and, even then, old ice still comes down from Victoria Strait. Freeze-up begins in the first week of October and turns the Gulf into a solid layer of ice by the end of the month.

Dease Strait, linking Queen Maud Gulf with Coronation Gulf, is about 100 miles in length with an eastern entrance of 12 miles and a western opening into Coronation Gulf of 38 miles. Dease Strait is covered by a solid sheet of first-year ice during the winter, which begins to fracture in the third week of July and usually leaves the Strait clear by the first week of August. Freeze-up begins in late September and the Strait is usually frozen solid by the third week of October.

Coronation Gulf, lying in an east–west direction, is over 100 miles long, with a western entrance into Dolphin and Union Strait 15 miles wide. The Gulf has numerous islands of various sizes throughout its length but is virtually free of islands along the south shore of Victoria Island. The Gulf is covered with a solid sheet of ice during winter, which begins to break up during the first week of July and normally clears by the end of the month. As a rule, freeze-up begins during the second week of October and consolidation of the ice cover is completed by December.

Dolphin and Union Strait measures about 90 miles long and ends in a western opening of 56 miles into Amundsen Gulf. There are several low islands and a number of reefs and shoals at the eastern end of the Strait, and the Sailing Directions recommend a specific route which should be followed closely. They specify that the least known depth within 0.5 mile of the recommended route is 13.2 metres, and it is further recommended that passage through the eastern part of the Strait be made only by day in good visibility.[7] The Strait is covered with a solid sheet of first-year ice during the winter and break-up normally begins during the first half of July. Freeze-up occurs about the middle of October.

Route 3A, through Peel Sound and James Ross Strait

This route is only a variation of Route 3, in that it passes on the east side of King William Island instead of the west side. This was the route followed by Roald Amundsen in 1903–6, when he became the first person to completely navigate the Northwest Passage. It is slightly longer than Route 3 and presents considerably more navigational difficulties. The maximum draft of vessels using this route has been 5 metres. The additional bodies of water traversed by this variation of Route 3 and which will now be described are the following: James Ross Strait, St Roch Basin, Rae Strait, Rasmussen Basin and Simpson Strait.

James Ross Strait lies in a general north–south direction between King William Island and Boothia Peninsula, and links Larsen Sound with St Roch Basin. It offers a 30-mile-wide passage, notwithstanding the presence of a number of large islands on its west side. However, because of several shoals at the south end of the Strait, the width of the navigable channel is reduced to 14 miles. This channel, plotted by the *Lindblad Explorer* in 1984, was found to have a minimum depth of 30 feet. Break-up normally begins by the end of July and clears by September, whilst freeze-up usually begins during the last week of September with a complete ice cover forming during the last week of October.

St Roch Basin completes the separation between King William Island and Boothia Peninsula and leads into Rae Strait to the south. The Basin is not completely free of islands and there is considerable shoaling when approaching Spence Bay on the west coast of Boothia Peninsula. Break-up in the Basin generally begins toward the end of July and freeze-up starts about the end of September or early October.

Rae Strait is a very short body of water about 13 miles wide, linking St Roch Basin with Rasmussen Gulf to the south. Reconnaissance soundings show an extensive shoal area in mid-channel, and the Strait does not appear suitable for deep draft navigation. Ice conditions are basically the same as those in St Roch Basin.

Rasmussen Basin, which is about 60 miles long, opens on to Simpson Strait to the west, with an entrance that is 13 miles wide. Except for a few islands in the approaches of Gjoa Haven at the southeast corner of King William Island, the Basin is clear. Depth soundings, however, have been very limited. A solid layer of ice covers the basin in winter and freeze-up usually begins during the first week of October and is completed by the end of the month. Break-up develops in the last half of July and the Basin is normally clear of ice by the middle of August.

Simpson Strait is a narrow passage, 45 miles in length and to the south of King William Island, with an eastern entrance about 7 miles wide, narrowing to about 2 miles and terminating in Queen Maud Gulf with an entrance about 8 miles wide. The Strait is obstructed by numerous islands and the least width of 2 miles is further reduced by the presence of small islands and shoals making the navigable channel as narrow as 0.15 mile in places. The channel has a least charted depth of 7.3 metres and Sailing Directions recommend a particular track through Storis Passage.[8] The Strait is obviously suitable for small ships only. Simpson Strait is covered with an even layer of ice in winter and freeze-up normally begins in the first week of October, the ice cover being completed by the last week of that month. Break-up normally develops in late July and, by the second week of August, the Strait is normally clear of ice.

Route 4, through Prince Regent Inlet

This fourth route is basically the same as Route 3 and was the one followed by Sargeant Henry Larsen, with his *St Roch* in 1940–2, in an easterly crossing. The only difference between Route 5 and Route 3 is that Route 3 follows the east side of Somerset Island through Prince Regent Inlet, instead of the west side through Peel Sound. The only other new body of water in this route is the short Bellot Strait at the southern tip of Somerset Island.

Prince Regent Inlet, separating Somerset Island from the northern part of Boothia Peninsula, runs from Lancaster Sound and joins with the Gulf of Boothia at the latitude of Bellot Strait. The inlet averages about 50 miles in width and is completely free of islands. Freeze-up normally begins in early October and occasionally includes the odd iceberg which has found its way through Lancaster Sound. Break-up generally begins at the northern end of the Inlet in late June and progresses slowly southward, reaching the latitude of Bellot Strait by the middle of August.

Bellot Strait is a short and very narrow passage between cliffy elevations of some 300 metres, offering generally deep water throughout the Strait but with a dangerous and extensive shoal area close to Magpie Rock where the depth is only about 80 metres. Sailing Directions specify that the Rock presents a formidable danger and two beacon ranges have been installed to lead ships through the narrow channel north of the Rock. The directions indicate how the Strait should be approached from the east and from the west and specify that passage should not be undertaken in reduced visibility.[9] The minimum depth on the range line is 22 metres.

Route 5, through Fury and Hecla Strait

This route could constitute an alternative to Lancaster Sound in order to reach Route 4 at Bellot Strait. Instead of proceeding into Davis Strait and Baffin Bay, a westbound ship would enter Hudson Strait and generally follow the southwest side of Baffin Island, then cross Fury and Hecla Strait and proceed into the Gulf of Boothia as far as Bellot Strait. The bodies of water traversed by the eastern portion of this route would be Hudson Strait, Foxe Channel, Foxe Basin, Fury and Hecla Strait and the Gulf of Boothia.

Hudson Strait, lying between the Ungava Peninsula and Baffin Island, is over 400 miles long and about 60 miles wide, and leads into Foxe Channel. The Strait has a few large islands at both entrances, as well as on either side, but the middle of the Strait is free of islands. Freeze-up begins in October and the Strait becomes unnavigable by the middle of November. The ice generally begins to break up at the end of June or early July, and the Strait is usually ice-free during August and September, except for the occasional iceberg, which has entered the Strait and sometimes persists throughout the summer.

Foxe Channel, lying between Southampton Island and Foxe Peninsula, is a deep body of water about 80 miles wide. A shoal is reported to lie in the middle of the Channel, reducing the depth to only 13 metres but there is ample room on both sides. Freeze-up begins during the latter part of October and is completed by early November. The ice in Foxe Channel

200 Canadian Archipelago and Northwest Passage

begins to break up in late June or early July and the Channel is usually free of ice by August.

Foxe Basin is a huge body of water between Melville Peninsula and Baffin Island, terminating at Fury and Hecla Strait. The northern section of the Basin in the approaches to Fury and Hecla Strait is characterized by an irregular bottom and several large islands with a number of dangerous shoals. Sailing Directions specify that particular caution must be exercised.[10] The ice conditions in Foxe Basin are essentially the same as in Foxe Channel. The ice is predominantly first-year ice but, due to the influence of wind and tide, there is considerable hummocking and ridging, with as many as 40 ridges per mile having been counted in the northern part of the Basin.

Fury and Hecla Strait is a passage about 105 miles long, joining Foxe Basin with the Gulf of Boothia. The Strait has a number of relatively large islands at its eastern entrance, and the narrowest part of the Strait, the Labrador Narrows, is 0.5 mile wide just south of a 3.9 m patch. The shallowest known depths in the Narrows vary from about 120 to 205 metres. West of Labrador Narrows the sounded channel widens to 2 miles, but an 11 m shoal restricts its width to 1.2 miles at one point.[11] The ice is made of local ice only and, because of the strength of the current, it rarely, if ever, forms in Labrador Narrows. Freeze-up usually develops in mid-October and break-up occurs in mid-July, with some ice remaining in August and September. So far, only a few ships have used this rather dangerous Strait.

The Gulf of Boothia is a continuation of Prince Regent Inlet to the north and is free of islands except for Crown Prince Frederick in the approaches to the western entrance to Fury and Hecla Strait. This island presents no navigational problems, however, and ships have ample room to the south of the island. The Gulf of Boothia is covered with consolidated ice in winter, a good proportion of which is made up of multi-year floes which have come down from Lancaster Sound. Freeze-up begins in early September and is completed by early October. Break-up begins somewhat later than in Prince Regent Inlet to the north and generally does not reach the latitude of Fury and Hecla Strait before mid-September.

Route 5A, through Fury and Hecla Strait and Prince Regent Inlet
A possible variation of Route 5 for a westbound ship is to proceed northward after crossing Fury and Hecla Strait, into Prince Regent Inlet

as far as Barrow Strait and thus join Route 1 or Route 2. This is the variation which Dome Petroleum is envisaging as an alternative route for its ships to reach its Beaufort Sea operations, most likely via Route No. 1. However, the shoals found in Fury and Hecla Strait in 1981 make it uncertain that this route could be used by deep draft vessels.

12.3. Summary

Of the five basic routes of the Northwest Passage just described, only two are presently suitable for navigation by deep draft ships of Arctic Class 7 or better. The water depths are adequate for such vessels up to 20 m draft in both Routes 1 and 2. Route 1, through Prince of Wales Strait, is the better of the two and is the primary route being considered by Dome Petroleum, because the prevailing ice conditions are less severe. Route 2, through M'Clure Strait, is usually clogged with polar ice and has seldom been completely navigated by a surface ship. Major leads, however, occasionally develop along the northern shore of the Strait, as well as along the west coast of Banks Island, which explains why Dome Petroleum has chosen it as a secondary route.

A third potential route for deep draft ships is Route 5A, through Fury and Hecla Strait. However, despite the fact that it was navigated three times by H.M.C.S. *Labrador* in 1956, the presence of shoals makes this Strait a doubtful alternative to Lancaster Sound for deep draft ships.

Routes 3 and 4 are adequate only for small ships, because of shoals and constrictions due to the presence of numerous islands. The maximum draft of vessels using Route 3A has been 5 metres.

13

Use of the Northwest Passage

The navigational regime applicable to a strait, which is not governed by treaty, depends on its use for international navigation.

13.1. Past use of the Northwest Passage[12]

Navigation in the Northwest Passage has taken place mainly for the purpose of exploration and adventure, before 1945, but also for security and economic reasons, since 1945.

Use of the Northwest Passage before 1945

Efforts to find a sea route to Cathay across the top of North America began with John Cabot in 1497 and lasted for nearly 400 years. Between Martin Frobisher's first expedition in 1576 and that of Sir John Franklin in 1845–7, some 40 expeditions, most of them British, sailed to the Arctic. The expeditions which contributed the most to the actual discovery of the Northwest Passage were those led by Lieutenant William E. Parry (1819–20, 1821–3, 1824–5), Sir John Ross (1829–33) and Sir John Franklin (1845–7). The routes followed by those expeditions are shown on Figure 21. During the 12-year period after the tragic loss of Franklin's expedition, over 75 search expeditons were sent to the Canadian Arctic. They resulted in the discovery and exploration of virtually all of the bodies of water which now constitute the main routes of the Northwest Passage (see Figure 13).

As already indicated in Chapter 8, Canada began sending expeditions to the Canadian Arctic shortly after the transfer of the Arctic Islands by Great Britian in 1880, in order to consolidate its title to the islands and exercise a certain degree of control over the waters of the Archipelago. Today, all of the routes of the Passage have been fully explored. The Norwegian explorer, Roald Amundsen, became the first to complete a crossing of the Northwest Passage in 1903–6. The main Canadian expeditions were those of A. P. Low in 1903–4, of Captain J. E. Bernier in 1906–7, 1908–9 (see Figure 21) and 1910–11, and of R.C.M.P. Sergeant Henry Larsen. The latter sailed from Vancouver to Halifax in 1940–2 and made the return trip in 1944 in a single season (see Figure 21).

> Fig. 21. Two of Canada's expeditions in the Northwest Passage: 1908–9 and 1940–4. Prepared for the author by the Canadian Hydrographic Service, Dept of Fisheries and Oceans, Ottawa. *Source*: from maps in A. Taylor, *Geographical Discovery and Exploration in the Queen Elizabeth Islands*, 1955, 115 and 130.

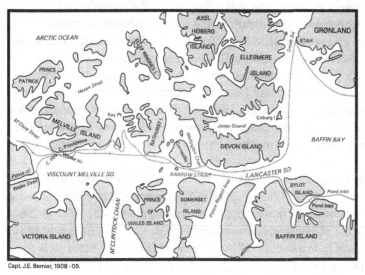

Capt. J.E. Bernier, 1908 - 09.

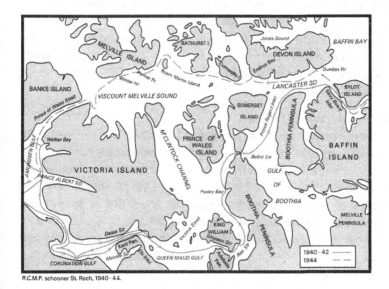

R.C.M.P. schooner St. Roch, 1940 - 44.

Use of the Northwest Passage since 1945
Aside from a few pleasure crossings, exploration and navigation activities in the Northwest Passage since World War II have been motivated primarily by security and economic reasons.

Use for security reasons
In 1954, the year before the conclusion of the Distant Early Warning line Agreement with the United States, Canada sent her naval icebreaker H.M.C.S. *Labrador* on a sovereignty and general survey mission through the Northwest Passage. The *Labrador*, under the command of Captain O. C. S. Robertson, completed her westward crossing by following Route 1 through Prince of Wales Strait. She thus became the first naval vessel and, thus far, the only one of Canadian registry, to complete the Passage.

In 1957, during the construction of the D.E.W. line installations, a squadron of three American Coast Guard icebreakers crossed the Northwest Passage from west to east, following Routes 3A and 4 through Bellot Strait and Prince Regent Inlet. The American ships were led through the dangerous Bellot Strait by the *Labrador* commanded by Captain T. C. Pullen.

During this period also, two American nuclear-powered submarines crossed the Northwest Passage, the U.S.S. *Seadragon* in 1960, following Route 2 in a westerly direction and, the other, the U.S.S. *Skate* in 1962, following the same Route in an easterly direction. The purpose of those crossings was to test the feasibility of submerged transits; these took place within the context of defence arrangements with Canada.

The only other crossing during this period, and only indirectly related to security, was that of the Canadian icebreaker *John A. Macdonald* in September 1967 which made a westerly crossing following Route 3. The *Macdonald* went to the rescue of the American icebreaker *Northwind*, which was beset in ice with a broken propeller some 500 miles north of Point Barrow.

Use for economic reasons
With the discovery of large oil reserves off Prudhoe Bay on the north slope of Alaska in 1968, oil companies began to consider the possibility of using the Northwest Passage to transport hydrocarbons to the American eastern seaboard. This resulted in the voyage of the *Manhattan* in 1969, the purpose of which was to determine the feasibility of year-round navigation. The *Manhattan* being accompanied by the Canadian icebreaker *John A. Macdonald* and the American icebreakers *Northwind* and *Staten Island*, the voyage resulted in a total of seven crossings of the Northwest Passage.

Table 3. *List of full transits of the Northwest Passage (1903 to 1985).* *

No.	Year	Name of ship	Registry	Route and direction	Type of ship	Nature of voyage
1	1903–6	Gjoa	Norway	3A west	herring boat	exploration
2	1940–2	St. Roch	Canada	3A and 4, east	schooner	patrol and exploration
3	1944	St. Roch	Canada	1, west	schooner	patrol and exploration
4	1954	HMCS Labrador	Canada	1, west	naval icebreaker	sovereignty and survey
5	1957	USCGS Storis	U.S.A.	3A and 4, east	icebreaker	hydrographic survey led through Bellot Strait by HMCS Labrador
6	1957	USCGS Spar	U.S.A.	3A and 4, east	icebreaker	exploration of submerged route, within Canada/U.S. defence arrangements
7	1957	USCGS Bramble	U.S.A.	3A and 4, east	icebreaker	exploration of submerged route, within Canada/U.S. defence arrangements
8	1960	USS Seadragon	U.S.A.	2, west	nuclear submarine	
9	1962	USS Skate	U.S.A.	2, east	nuclear submarine	
10	1967	CCGS John A. Macdonald	Canada	3, west	icebreaker	assist CCGS Camsell in MacKenzie Bay and USCGS Northwind n. of Pt. Barrow
11	1969	S/T Manhattan	U.S.A.	1, west	tanker	test Passage for large tanker, with Can. Navy Capt. T. C. Pullen aboard
12	1969	CCGS John A. Macdonald	Canada	1, west	icebreaker	escort and assist Manhattan
13	1969	S/T Manhattan	U.S.A.	1, east	tanker	return voyage
14	1969	USCGS Northwind	U.S.A.	3, east	icebreaker	rendezvous and support Manhattan
15	1969	CCGS John A. Macdonald	Canada	1, east	icebreaker	support Manhattan
16	1969	USCGS Staten Island	U.S.A.	1, east	icebreaker	support Manhattan

continued

Table 3 (*continued*).

No.	Year	Name of ship	Registry	Route and direction	Type of ship	Nature of voyage
17	1969	USSGS *Northwind*	U.S.A.	3, west	icebreaker	return to Seattle
18	1970	CSS *Baffin*	Canada	1, east	survey ship	survey
19	1970	CSS *Hudson*	Canada	1, east	survey ship	survey (first circum. of Americas)
20	1975	CCGS *Skidegate*	Canada	3A, and 4, east	icebreaker	redeployed to east coast
21	1975	CCGS *John A. Macdonald*	Canada	3, west	icebreaker	assist CCGS *Camsell* and *Beaver Mackenzie* in Beaufort Sea
22	1975	CCGS *John A. Macdonald*	Canada	3, east	icebreaker	return to east coast
23	1975	MV *Pandora II*	Canada	3 and 5, east	survey ship	survey
24	1975	MV *Theta*	Canada	3 and 5, east	survey ship	survey
25	1976	*Canmar Explorer II*	Canada	3, west	drill ship	drill for Dome in Beaufort Sea
26	1976	CCGS *J. E. Bernier*	Canada	3, east	icebreaker	redeployed to east coast
27	1977	*Williwaw*	Netherlands	3A, west	yacht	adventure
28	1977–8	*J. E. Bernier II*	Canada	3A, west	yacht	adventure
29	1978	CCGS *Pierre Radisson*	Canada	1, east	icebreaker	redeployed to east coast
30	1978	CCGS *John A. Macdonald*	Canada	1, west	icebreaker	charter to Dome in Beaufort Sea
31	1979	CCGS *John A. Macdonald*	Canada	1, east	icebreaker	return from charter
32	1979	MV *Kigoriak*	Canada	1, west	icebreaker	support for Dome in Beaufort Sea.
33	1979	CCGS *Louis St-Laurent*	Canada	1, west	icebreaker	assist CCGS *Franklin* in Visc. Melville Sd.
34	1980	CCGS *J. E. Bernier*	Canada	3A, east	icebreaker	returning to Quebec from Beaufort Sea
35	1980	MV *Pandora II*	Canada	3A, east	survey	redeployed to east coast
36	1981	CSS *Hudson*	Canada	3, east	survey	return to east coast from survey off west coast
37	1981	*Morgan Stanley*	Canada**	3A, east	Boston whaler	adventure

206

38	1979–82	*Mermaid*	Japan	3A and 4, west	yacht	adventure
39	1983	*Arctic Shiko*	Canada	3, east	tug	supply
40	1983	*Polar Circle*	Canada	3A, east	survey ship	survey
41	1984	*M/S Lindblad Explorer*	Bahamian (Swedish owned)	3A, west	cruise ship	pleasure trip St John's to Yokohama; permission requested
42	1985	*M/S World Discoverer*	Liberian	3, east	cruise ship	pleasure trip Nome to Halifax; permission requested
43	1985	USCGS *Polar Sea*	U.S.A.	1, west	icebreaker	redeployed from Thule to Chukchi Sea for survey work; permission not requested
44	1985	CCGS *John A. Macdonald*	Canada	3, west	icebreaker	assist CCGS *Camsell* in Beaufort Sea
45	1985	CCGS *John A. Macdonald*	Canada	3, east	icebreaker	return to east coast

* The writer is greatly indebted to Captain T. C. Pullen, R.C.N. who compiled a list of full transits on which this one is based. The Northwest Passage here is limited to the constricted waters of the Canadian Arctic Archipelago between Baffin Bay and Beaufort Sea.

** Ship bought in Vancouver just prior to crossing and presumably of Canadian registry; see R. Fiennes, *To the Ends of the Earth*, at 158 (1983).

Canada's exploration activities for offshore hydrocarbon resources in the Beaufort Sea also resulted in a number of crossings of the Northwest Passage. Indeed, except for four adventure crossings, two pleasure cruises and a U.S. icebreaker redeployment,[13] the 21 other crossings related to those exploration activities and were all completed by Canadian ships. These were either survey ships, icebreakers, drill ships or tugs. The name of the ship, the registry, the route and direction followed, the type of ship and the nature of the voyage are all shown on a list of the full transits of the Northwest Passage from 1903 to 1985 (see Table 3).

In addition to the above, Canadian Coast Guard icebreakers have made a number of partial transits of the Northwest Passage, particularly in the last eight or nine years, with the development of mining activities in the Canadian Arctic Islands. From 1977 to 1985 inclusively, a total of 199 ships made partial transits of the Northwest Passage and 159 of them were of Canadian registry (see Table 4).

These included icebreakers, survey ships, fuel tankers, general cargo ships, bulk carriers, tugs and drill ships. Most of the ships, 172 out of 199, entered the passage through Lancaster Sound. Virtually all of the Canadian ships penetrated at least half-way across the passage and 23 of them as far as the western half of Viscount Melville Sound. Since 1977, when the ship-reporting system NORDREG was established, ship movements take place under the supervision and control of the Canadian Coast Guard.

13.2.   Future use of the Northwest Passage

With the discovery of substantial hydrocarbon resources in areas adjacent to the Northwest Passage and the advances in technology for their exploitation and transportation, it is only a question of time before the Northwest Passage is used for commercial navigation. This section examines the approximate extent of the resource potential and the projected marine traffic in the Northwest Passage.

Resource potential in areas adjacent to the Northwest Passage

In addition to a few lead and zinc mines, such as the ones which are already operating on Baffin Island and Little Cornwallis Island, considerable hydrocarbon resources are located in the Alaskan North Slope and Mackenzie Delta/Beaufort Sea regions, as well as in the Canadian Arctic Islands themselves.

Alaska North Slope/Beaufort Sea[14]

The Prudhoe Bay field, discovered in 1968 on the central north slope of Alaska, was brought into production in 1977 upon completion of the

Trans Alaska Pipeline System. The recoverable reserves of Prudhoe Bay are estimated to be about 10 billion barrels of oil and currently produce approximately 1.5 million barrels per day of crude oil. The other field west of Prudhoe is Kuparuk, which is now producing over 100,000 barrels of oil per day and has an estimated recoverable reserve of 1.5 billion barrels of oil.

Although there has yet to be any offshore production in the Beaufort Sea off the Alaskan coast, numerous exploratory wells have been drilled and there has been a considerable increase in leasing activity in recent

Table 4. *Partial transits of Northwest Passage (1977–85).* *

		Entry from east		Entry from west	
Route	Water body	Canadian	Foreign	Canadian	Foreign
Route 1	Lancaster Sound	135	37[1]	0	0
	Barrow Strait	98	20[2]	0	0
	Visc. Melville Sound	23	7[3]	1	0
	Prince of Wales Strait	1	0	1	0
	Amundsen Gulf	0	0	23	2[4]
Route 2	M'Clure Strait	0	0	0	0
Route 3	Peel Sound	4	0	0	0
	Franklin Strait	3	0	0	0
	Victoria Strait	3	0	0	0
	Queen Maud Sound	0	0	19	2[4]
	Coronation Gulf	0	0	23	2[4]
Route 3A	James Ross Strait	0	0	4	0
	Rae Strait	0	0	16	2[4]
	Simpson Strait	0	0	16	2[4]
Route 4	Prince Regent Inlet	1	0	0	0
	Bellot Strait	0	0	0	0
Route 5	Hudson Strait	150	131[5]	0	0
	Foxe Basin	39	0	0	0
	Fury and Hecla Strait	0	0	0	0
	Gulf of Boothia	1[6]	0	0	0
Route 5A	This route has not been used.				

1. Ships of Finland (8), France (2), Japan (1), Liberia (16), Panama (1), Sweden (8) and West Germany (1).
2. Ships of Finland (8), Japan (1), Liberia (6) and Sweden (3).
3. Ships of Finland (7).
4. Ships of United States.
5. No transits by foreign ships north of Foxe Channel.
6. Entry via Prince Regent Inlet and Lancaster Sound.

*The author is indebted to Captain J. Y. Clarke, Director General, Fleet Systems, Canadian Coast Guard, for the compilation of the list of partial transits on which this summary is based.

years. More drilling is scheduled to take place, not only in the Beaufort Sea, but in the Bering Sea as well. In addition, considerable hydrocarbon reserves also exist in Alaska itself, more specifically, in the National Petroleum Reserve Area and the Arctic National Wildlife Refuge.

Estimates of recoverable resources of Alaskan oil and gas potential have been made both by the United States Geological Survey and the National Petroleum Council. Both of their estimates are very close, with those of the Geological Survey being slightly lower. The latter's estimates show a recoverable potential of 19 billion barrels of oil and 102 trillion cubic feet of gas.

The throughput of the 800-mile pipeline which is presently being used to transport oil to Valdez could be increased by as much as a third, but if the exploratory efforts result in a sizeable additional production, delivery to market will have to be made either through a second pipeline or by a tanker system. It is interesting to note in this regard that the Committee on Maritime Services to Support Polar Resource Development recommended the study of the possible use of the Northwest Passage in its 1981 report.[15]

Mackenzie Delta/Beaufort Sea.[16]
The sedimentary basin of the Mackenzie Delta/Beaufort Sea comprises an area of about 420,000 square kilometres, only a quarter of which has so far been explored. Estimates of its oil and gas potential have varied considerably between the ones made by developers and those of the Geological Survey of Canada. The latter's estimates of recoverable reseves are 9.4 billion barrels of oil and 112 trillion cubic feet of gas.

Although the 1983 projection by Dome Petroleum 1983 that production in the Beaufort Sea could begin as early as 1989 was probably too optimistic, future production does appear to be a certainty. The present Government, as the previous one, is committed to the development of northern hydrocarbon resources and its policy statement of October 30, 1985 specified that 'oil reserves discovered in the Beaufort Sea are . . . nearing the threshold required to justify development'.[17] Details of production scenarios from the Beaufort Sea area are outlined in the Environmental Impact Statement prepared by Dome Petroleum, Esso Resources Canada and Gulf Canada of 1982[18] and were reviewed by the Special Committee of the Senate on the Northern Pipeline in 1983. Dome Petroleum estimates that oil production would be commercially feasible at a threshold reserve level of only 700 million barrels of oil if the tanker mode of transportation is used, whereas a pipeline system would require some 2500 million barrels.[19]

Canadian Arctic islands.[20]

The hydrocarbon resources in the Canadian Arctic Islands are located mainly on the north side of the Northwest Passage in the Queen Elizabeth Group. Those resources are estimated by the Geological Survey of Canada to be in the order of 4.3 billion barrels of recoverable oil and 87 trillion cubic feet of gas. The principal operator in the area, Panarctic Oils, had hoped to deliver gas to southern markets by tankers as early as 1986, but the Arctic Pilot Project has now been deferred. However, Panarctic shipped 100,000 barrels of light crude oil from its Bent Horn oilfield on Cameron Island in 1985 and plans to ship the same quantity in 1986 and 1987. At the beginning of the second phase of its project in 1988, the plan is to increase to 300,000 barrels a year and eventually to about 50,000 barrels a day with large ice-breaking tankers.[21] The implementation of the Bent Horn project will be most valuable in preparing for tanker transportation from the Beaufort Sea under conditions which conform with the recommendations of the Beaufort Sea Environmental Assessment Panel.

Projected marine traffic in the Northwest Passage

There exist two major modes of transporting oil and gas resources from the Arctic regions adjacent to the Northwest Passage: pipelines and tankers. What follows is a brief comparative examination of those two modes, along with a tentative projection of marine traffic which might take place in the foreseeable future in the Northwest Passage.

Pipeline mode of transportation.[22]

In addition to the existing Trans Alaska Pipeline, four pipeline proposals have been made for the transportation of Arctic hydrocarbons to southern markets. The *Alaska Highway Pipeline*, a joint Canadian/American project, would bring natural gas from the Alaskan north slope area over a distance of some 7700 kilometres, through southern Yukon, northern British Columbia and Alberta, to southern markets. It is designed to connect with the proposed 1200-kilometre Dempster Lateral Line near Whitehorse to feed gas from the Mackenzie Delta area into the main pipeline. A number of problems, including financial ones, have plagued the project and, although some of them have been alleviated, it appears certain that the completion of the pipeline by the scheduled date of 1999 is not going to be met. Indeed, it is possible that the pipeline might never be built.

The *Dempster Lateral Line* would bring up to 1.2 billion cubic feet of natural gas per day to the Alaska Highway Pipeline and the gas transported by this line would be destined primarily for Canadian markets. The completion of this line, of course, is contingent upon the building of the Alaska Highway Line as well as a timely completion of a third line, the Polar Gas 'Y' Pipeline.

The *Polar Gas 'Y' Pipeline project*, formed in 1972, originally sought to develop a mode of transporting natural gas from the discoveries made by Panarctic Oils in the Arctic Islands. Once it was determined that a large diameter pipeline would probably be required, alternative plans were considered and, in 1980, the concept of a Y configuration pipeline linking discoveries in the Arctic Islands with those of the Mackenzie Delta/ Beaufort Sea was devised. Although originally slated for completion by 1990, there are a number of remaining uncertainties and contingencies so that this third proposal might not materialize.

The fourth proposal is for the building of the *Beaufort Sea–Edmonton Pipeline*, as an alternative to a tanker system, which would carry oil about 2250 kilometres from Richards Island off the Beaufort Sea, along the Mackenzie Valley to Norman Wells, and continue south past Fort Simpson as far as Zama, northwest of Edmonton. The major proponent of this pipeline is Esso Resources of Canada, one of the three developers in the Beaufort Sea. Esso has already completed the construction of a pipeline from the existing oil field at Norman Wells to Zama.[23] Proposals are still being developed to extend the pipeline northward from Norman Wells to the Beaufort Sea.

Tanker mode of transportation

Three Canadian and at least five American proposals have been made to transport hydrocarbon resources through the Northwest Passage. One of the Canadian proposals advanced by *Dome Petroleum* involves the transportation of Beaufort Sea oil by Arctic Class 10 tankers. These 200,000 DWT tankers, some 390 metres long and 52 metres wide, would have a double shell and would be capable of year-round navigation. The primary shipping corridor would be Route No. 1 through the Prince of Wales Strait, but Route No. 2 through M'Clure could also be used as an alternative route. A second Canadian proposal is the *Arctic Pilot Project* which was designed to transport natural gas produced from northern Melville Island, liquefied and stored on the south coast of the island and then transported through the Northwest Passage in Arctic Class 7 LNG tankers of 140,000 cubic metre capacity. These tankers would be 372 metres long and 43 metres wide. It now appears that this project of Petro Canada has been delayed indefinitely and might never come on stream.

A more recent and modest project which has already shown signs of success is that of Panarctic Oils which involves the development of the

Bent Horn Oil Field, on Cameron Island, north of its Rae Point exploration base. Panarctic shipped 100,000 barrels of light crude oil in 1985 and
plans to ship eventually about 50,000 barrels a day, using ice-breaking
tankers. This is the kind of small scale project which fits well within the
parameters of Canada's strategy for northern hydrocarbon development[24] and has a much greater chance of materializing. Under this
strategy, the Government will encourage initial production on a small
demonstration basis, commensurate with proven commercial reserves,
and will support incremental expansion as need and capability are
demonstrated.[25]

The main American proposals have been made by the U.S. Maritime
Administration, Sea Train Lines, Globtik Tankers and General
Dynamics. The proposals made by the first two proponents envisage the
use of large ships varying from 225,000 to 350,000 DWT. These huge
tankers would transport oil, and possibly gas eventually, to the American
east coast through the Northwest Passage. The proposals made by
General Dynamics foresee the use of submarine tankers which would be
able to go through the Northwest Passage unaffected by surface ice and
weather conditions. This type of huge submarine tanker would appear to
be somewhat futuristic at this time and perhaps also rather unrealistic,
when one considers that their estimated safe operating depth is 150
metres and mid-channel depths in Barrow Strait are as shallow as 22
metres.

It is very difficult to predict which of the two will be chosen as the basic
mode of transportation, but it does appear that the inherent flexibility of
the shipping mode represents a definite advantage over the pipeline
system. Not only does it facilitate the use of small demonstration projects,
but it permits an incremental expansion as reserves, need and safety of
transportation are demonstrated. On the other hand, a pipeline system of
over 2000 kilometres through arctic and subarctic terrain may cost in the
neighbourhood of 12 billion dollars and would take at least ten years of
lead time to plan and finance.[26] Consequently, it is not surprising that the
Canadian Senate Special Committee on the Northern Pipeline concluded
by recommending 'that transport of hydrocarbons from the arctic region
commence by tanker on a small scale and that consideration be given to
various combinations of tanker and/or pipeline systems as other factors
warrant'.[27]

Tentative projection of marine traffic
There has been a number of fairly significant indications that Canada is
preparing itself to meet the demands of substantial marine traffic in the
Northwest Passage. Not only did it announce its planning strategy for a

phased approach to northern hydrocarbon development in June 1982, but in the same year Transport Canada adopted an Arctic Marine Service Policy to 'provide for an appropriate level of services and regulation in support of marine transportation and related activities in the Canadian Arctic'.[28] In addition, and also in 1982, the same Department prepared a number of possible scenarios of marine traffic in the Northwest Passage, as part of its position statement to the Beaufort Environmental Assessment Panel. These scenarios, relating to oil and gas transportation, range from those of low or no activity to those of high activity. The latter scenario would involve oil and gas shipments by icebreaking tankers commencing as early as 1986 for gas from the Arctic Islands and 1987 for oil from the Beaufort Sea and would then escalate fairly rapidly after that.

Table 5. *Assumed most demanding scenario forecast of Arctic marine traffic levels*[29]
(Number of one-way ship transits)

Oil/gas scenario, year-round operation	1980	1985	1986	1987	1988	1990	1992	1993	1994	1995
Beaufort Sea (Oil)	—	—	—	24	48	72	96	120	144	168
Arctic Islands (Oil)	—	—	—	—	—	—	—	30	60	60
Labrador (Oil)	—	—	—	—	—	—	—	—	—	36
Arctic Islands (Gas)	—	—	30	60	60	90	120	150	180	120
Alaska	—	—	—	—	—	—	—	24	48	72

Note: Figures assumed as of September 10, 1980.

It is now quite evident that this most demanding scenario will not materialize, but it is nevertheless given here as evidence of the seriousness with which the Government considers the likelihood of future marine traffic in the Passage. Whilst it is now evident that the commencement dates are inaccurate, it is still possible to envisage the shipment of oil and gas in the 1990s.

Regardless of the exact date of the beginning of marine traffic and of its precise extent, it appears quite certain that it will become a reality. This, of course, raises the question of government policy with respect to both domestic and foreign shipping, as ships of foreign registry are bound to be used to an appreciable extent. Consequently, the Department of Transport has been preparing an Arctic Marine Shipping Policy covering both types of shipping. In addition, the use of the Northwest Passage for international navigation raises the question of its legal status, both present and future.

14

Legal status of the Northwest Passage

Canada's position as to the legal status of the Northwest Passage was expressed in 1975 by the Secretary of State for External Affairs Allan MacEachen. Appearing before the Standing Committee for External Affairs and National Defence in May of that year, Mr MacEachen stated:

> As Canada's Northwest Passage is not used for international navigation and since Arctic Waters are considered by Canada as being *internal waters*, the régime of transit passage does not apply to the Arctic.[30]

This official position of Canada on the Northwest Passage is clear enough as far as it goes, but it does leave at least three questions unanswered. These questions will now be examined and are the following: 1) When may a strait be considered as being used for international navigation; 2) What is the legal status of the Northwest Passage itself; and 3) What would be the legal status of the Northwest Passage if it were used for international navigation?

In spite of the conclusion in Chapter 11 that Canada had validly established straight baselines around the Arctic Archipelago, the questions just formulated must still be addressed.

14.1. Definition and classification of international straits

At the Third Law of the Sea Conference, the question of international straits was dealt with as a distinct subject, separate from that of the territorial sea. This represents a considerable improvement over the approach taken at the First Law of the Sea Conference and the consequent provisions found in the Territorial Sea Convention of 1958. After much difficulty, the Conference did reach a compromise as to the nature and scope of the right of passage which would be applicable to 'straits used for international navigation'. However, it did not succeed in formulating a definition of such straits, but merely specified the categories of straits to which the new right of passage would apply. Consequently, one has to rely on existing customary international law for a definition of an international strait.

Definition of an international strait

As in numerous other instances of customary law, there is some uncertainty as to what is 'a strait used for international navigation'. Attempts at codifying the customary law applicable to straits were made by a number of international bodies, such as the Institut de Droit International, the International Law Association and The Hague Codification Conference, but without much success on the important question of the degree of use required for a strait to be considered international.[31] Fortunately, the International Court did address this point in the *Corfu Channel Case* of 1949 and this is still the only international decision on the question. The codification attempts having taken place before that decision, and none having been made since, aside from those of the Third Law of the Sea Conference, the decision represents the best evidence of what the state of customary law is on the legal requirement of an international strait. The International Court confirmed that an international strait had to meet two criteria, one pertaining to geography and the other, to the use or function of the strait.

Geographic criterion

As a rule, whenever there is an overlap of territorial waters in the natural passage between adjacent land masses joining two parts of the high seas, the necessary geographic element is present for the passage to be called a territorial or legal strait, and a special legal regime becomes applicable. If there is no such overlap, and a strip of high seas remains throughout the strait, the principle of the freedom of the high seas continues to apply.

Although the above description, representing the traditional definition of a legal strait, applies to most straits, three qualifications must be made. Those qualifications relate to the nature of the waters being joined, the width of the overlap of territorial waters and the pockets of high seas which might be found in the strait. With respect to the bodies of sea water being joined, it is no longer necessary since 1958 for the strait to join two parts of the high seas, but it is sufficient for it to join a part of the high seas with the territorial sea of a foreign State. And with the incorporation of the exclusive economic zone in the 1982 Convention, a strait may join two parts of exclusive economic zones. The second qualification relating to the overlap is that, since a 12-mile territorial sea is now permitted in international law, a legal strait means one which is 24 miles or less in width. In addition, under the 1982 Convention, a strait of more than 24 miles and used for international navigation might be considered a legal strait, if the high seas route in the middle is not suitable or convenient for navigation.[32] The third qualification relates to the enclaves or pockets of high seas which might be left in a strait where the territorial waters on either side do not reach. The Convention does not cover this situation, but there seems to be a consensus of opinion that these pockets of high seas should be assimilated to territorial waters.[33] The result of this view is that, if there is only one littoral State, the pockets of high seas in the strait are assimilated to the territorial sea of that State; otherwise, the enclaves will be divided along the median line between the littoral States as additions to their regular territorial waters.

Functional criterion
The required degree of use for international navigation is much more difficult to apply than the geographic criterion just discussed. Both the 1958 Territorial Sea Convention and the 1982 Convention are silent on the important question of how the necessary degree of use for international navigation is to be determined. This issue is a complex one, since it is necessary to determine what factors are relevant and what their relative importance should be.

It is probably the Danish jurist, Erik Bruël, who has studied this issue in the most exhaustive manner. In his definitive two-volume treatise entitled *International Straits*, published in 1947, he suggested that the question ought to be determined on the basis of such facts as 'the number of ships passing through the strait, their total tonnage, the aggregate value of their cargoes, the average size of the ship and especially whether they are distributed among a greater or smaller number of nations – all of which seem to give good guidance, no single factor, however, being decisive . . .'.[34]

Having enumerated the relevant factors to take into account, Bruel concludes that only those straits that are of considerable importance to international maritime commerce enjoy the peculiar legal position accorded to international straits. By the qualification 'international', he writes, 'is simply meant that the interest attached *to the use of* these straits is world wide.'[35] He adds that the number of straits endowed with this quality is limited and that the special conservatism of shipping in the use of routes makes it reasonably clear what straits are considered important.

Bruel further indicates that a number of qualifying expressions have been developed over the years to point to what are considered to be international straits. The expressions mentioned are 'routes maritimes indispensables', 'routes maritimes nécessaires à la navigation', 'grandes routes maritimes', 'passage habituel', 'international highways', 'highways for international traffic' and 'natural traffic routes'.[36]

The above represented quite accurately the state of the law when, in 1949, the International Court had to pronounce itself on whether or not the North Corfu Channel was an international strait. The Court arrived at the conclusion that 'the North Corfu Channel should be considered as belonging to the class of international highways through which passage cannot be prohibited by a coastal State in time of peace.'[37] The judgment had to deal with the arguments of the United Kingdom which put the emphasis on the geographic criterion, and those of Albania which insisted on the functional one.

The United Kingdom contended that 'the character of the channel as an international route depends on the fact that it connects two parts of the open sea and is useful to navigation, not on the volume of traffic passing through it'.[38] It adduced evidence showing that, during the period from April 1, 1936 to December 31, 1937 (one year and nine months), some 2884 ships of seven different nationalities had put in at the Port of Corfu after passing or just before passing through the Channel. The Court was also informed that the British Navy had regularly used the Channel for some 80 years or more as had the navies of other States as well.[39]

Albania, on the other hand, contended that the Corfu Channel had very little importance for international navigation and was not a highway *(grande route)* but only a byroad *(une voie latérale et secondaire)*. It contended that this secondary route was practically reserved to local navigation and cabotage and, therefore, did not serve for passage properly so-called.[40]

In answering the question 'whether the test is to be found in the volume of traffic passing through the Strait or in its greater or lesser importance for international navigation', the Court stated that 'the decisive criterion is rather its geographical situation as connecting two parts of the high seas and the fact of its being used for international navigation'.[41] The French text of the judgment, which is the authoritative one, reads: '*Le critère décisif paraît plutôt devoir être tiré de la situation géographique du Détroit, en tant que ce dernier met en communication deux parties de haute mer, ainsi que du fait que le Détroit est utilisé aux fins de la navigation internationale*'.[42] It would appear, therefore, that in the French text, a little less importance is given to the requirement that the strait be used for international navigation.

The Court held that the fact that the North Corfu Channel was not a necessary route between two parts of the high seas, but only an alternative passage, could not be decisive. What was more important was that it had been a 'useful route for international maritime traffic'.[43] The evidence showed that it had been a very useful route for the flags of seven States: Greece, Italy, Romania, Yugoslavia, France, Albania and the United Kingdom. The 2884 crossings, counted during a 21-month period, covered only the ships which had put in port and had been visited by customs. It did not include the large number of vessels which went through the strait without calling at the Port of Corfu. In other words, the actual use of the North Corfu Channel had been quite considerable.

With this decision in mind, it is not surprising that the International Law Commission, in its Draft Convention in preparation for the 1958 Law of the Sea Conference intended to codify the existing customary law on the point, confined the right of non-suspendable innocent passage to straits '*normally* used for international navigation between two parts of the high seas'.[44] The qualifying adverb 'normally', however, was not retained in the Territorial Sea Convention of 1958 and is absent from the 1982 Convention. Consequently, there is no indication whatever in conventional law as to what degree of use for international navigation is required before a strait can qualify as an international strait. It is obvious that the actual use has to be considerable, although it might not have to be as extensive as it was in the *Corfu Channel Case*. As pointed out by Fitzmaurice, 'in the *Corfu Case*, the Court clearly thought that the Channel was used to an *important* extent'.[45]

As for the factors relating to use, the Court does not seem to have considered two of the factors mentioned by Bruel, namely, the total tonnage and the aggregate value of their cargoes. It did take into account, however, the two other factors, namely, the number of ships passing through the strait and the number of flag States represented by those ships. It also appears to have considered the evidence presented by the United Kingdom 'that the British Navy has regularly used this Channel for eighty years or more, and that it has also been used by the navies of other States'.[46]

Having regard to the above, perhaps the only effect of deleting the qualifying adverb 'normally' from the I.L.C. draft, probably intended to cover such regular use as was made by the British Navy for some 80 years, was to remove the necessity of such regular use for that long period of time. In other words, the simple phrase 'used for international navigation' would represent a somewhat modified codification of the *Corfu Channel*

Case principle, in that it would not include the regular use by the British Navy but only the 2884 crossings or more made by the ships of seven States during a 21-month period. Regardless of the precise reasons for the insertion and the removal of the word 'normally', it seems clear that, before a strait may be considered international, proof must be made that it has *a history as a useful route for international maritime traffic.*

This conclusion is in accord with the interpretation generally given to the Court's judgment. For instance, Professor Baxter writes that 'international waterways must be considered to be those rivers, canals, and straits which are *used to a substantial extent* by the commercial shipping or warships belonging to states other than the riparian nation or nations'.[47] As to the criteria applied by the International Court in the *Corfu Channel Case*, Professor Baxter concludes that 'the test applied by the Court lays more emphasis on the practices of shipping than on geographic necessities'.[48]

The importance of the functional element is also emphasized by Professor O'Connell in the following passage:

> When it is said, then, that a strait in law is a passage of territorial sea linking two areas of high sea this is not to be taken literally, but rather construed as meaning a passage which ordinarily carries the bulk of international traffic not destined for ports on the relevant coastlines. *The test of what is a strait*, unlike the test of what is a bay, *is not so much geographical, therefore, as functional.*'[49]

Professor O'Connell re-emphasized the importance of the functional criterion in his most authoritative study, *The International Law of the Sea*, published posthumously in 1982. In his opinion, the *Corfu Channel Case* established 'that not all straits linking two parts of the high seas are international straits, but *only those which are important as communication links*'.[50] He concluded that 'the real significance of the Case is its treatment of the qualities that give a strait this importance'. In his view, 'it is clear that it need not be the only or an indispensable or a necessary avenue, but equally clear that mere *potential utility is insufficient*'.[51]

On the basis of the above analysis, as to the functional criterion in the determination of what is an international strait, the following propositions may be formulated:

1. actual use of a strait, by opposition to mere potential use, is necessary;
2. the strait does not have to constitute a necessary route for international navigation and may be only an alternative one;

3. the strait must have a history as a useful route for international maritime traffic and episodic or infrequent transit is insufficient;
4. the sufficiency of the use is determined mainly, although not exclusively, by reference to two factors, namely, the number of transits and the number of flags represented; and
5. the numbers of transits and flags should normally be substantial, but the location of the strait and other relevant circumstances might render lower numbers sufficient.

Classification of international straits

After considerable discussion as to the type of passage which should be permitted in international straits, a consensus was reached at the Third Law of the Sea Conference, to provide for at least five categories of international straits and four different legal regimes. These categories of straits do not affect those already governed by longstanding conventions. What follows is a brief description of the five categories of international straits with their attendant legal regime (see Figure 22).

1. International straits, with a route of high seas (or exclusive economic zone) of 'similar convenience' within them, will be subject to the freedoms of navigation and overflight applicable to the high seas and exclusive economic zone.[52] The determining criterion of 'similar convenience' refers to 'navigational and hydrographical characteristics'.[53] Such straits join parts of high seas or exclusive economic zones.
2. International straits, with a route of high seas (or exclusive economic zone) not of 'similar convenience', will be governed by the right of transit passage.[54] This category of straits is provided for by implication in Article 36, which states that 'this Part does not apply to a strait used for international navigation if there exists through the strait a route through the high seas or through an exclusive economic zone of similar convenience'. If the condition for the exclusion of such straits is not met, they are included in the application of Part 3 on international straits and subject to the right of transit passage. These straits join parts of high seas or exclusive economic zones.
3. International straits, with a high seas (or exclusive economic zone) route of 'similar convenience' seaward of an island (which forms the strait) of the coastal State, is subject to the right of non-suspendable innocent passage.[55] The question of 'similar convenience' is always in relation to navigational and hydro-

Fig. 22. International Straits in 1982 Law of the Sea Convention. Prepared for the author by John Cooper, Technical Advisor to Dept of External Affairs, Ottawa.

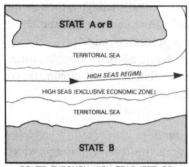

ROUTE THROUGH HIGH SEAS (EEZ) OF
SIMILAR CONVENIENCE – ARTICLE 36

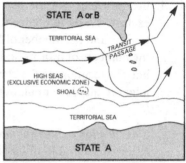

ROUTE THROUGH HIGH SEAS (EEZ) NOT OF
SIMILAR CONVENIENCE – ARTICLE 36

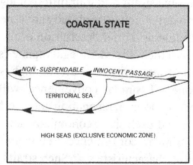

ROUTE THROUGH HIGH SEAS (EEZ) OF
SIMILAR CONVENIENCE SEAWARD OF ISLAND
– ARTICLES 38 (1) AND 45 (1) (A)

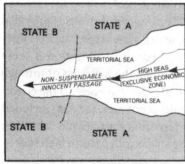

ROUTE JOINING HIGH SEAS (EEZ) WITH
TERRITORIAL SEA OF FOREIGN STATE
– ARTICLE 45 (1) (B)

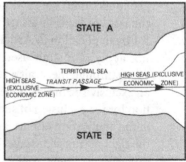

ROUTE JOINING HIGH SEAS (EEZ) WITH
HIGH SEAS (EEZ) – ARTICLES 37 AND 38 (2)

graphical factors. These straits join parts of high seas or exclusive economic zones.

4. International straits, joining a part of the high seas (or an exclusive economic zone) with the territorial sea of a foreign State, are governed by the traditional right of non-suspendable innocent passage.[56]
5. International straits, joining a part of the high seas (or an exclusive economic zone) with another part of the high seas (or exclusive economic zone) and not included in the previous categories, are subject to the right of transit passage.

The categories of international straits just outlined would seem to be exhaustive, except for the possibility that a strait comprised exclusively of internal waters might be used for international navigation. Two different legal situations might result, depending on the legal basis for the status of the waters. If the waters are internal because of the establishment of straight baselines, they are assimilated to territorial waters for the purposes of passage and either transit passage or non-suspendable innocent passage would apply as already discussed. If the waters are internal by reason of historic title, no right of passage applies and the strait would normally not be capable of becoming an international strait. The only possible exception to the latter would result from the coastal State permitting foreign use of the strait, without adequate conditions attached. The right of passage would then result from the acquisition of a servitude, rather than from the status of the waters *per se*.

14.2.   Legal status of the Northwest Passage

Three main questions must be addressed in the examination of the legal status of the Northwest Passage: (1) Is it an international strait? (2) Was there a right of passage before the straight baselines? and (3) Is there a right of passage now?

Is the Northwest Passage an international strait?

Applying the definition of an international strait discussed in the previous section, the Northwest Passage must meet two criteria, one, geographic and the other functional, before it can be classified as an international strait.

The *geographic criterion* is met without difficulty in that it links two parts of the high seas. Indeed, the eastern end of the Passage leads to Baffin Bay, Davis Strait, the Labrador Sea and the Atlantic Ocean, whereas the western end leads to the Beaufort Sea, the Chukchi Sea, the Bering Strait and the Pacific Ocean. It has occasionally been suggested that the Beaufort Sea, which is part of the Arctic Ocean, should not be considered as high seas because of the presence of ice. However, such a suggestion seems to leave aside the fact that the pack ice, which is not an ice cap, is comprised of ice floes in constant motion and does not extend to the continental coast of Canada and Alaska for about three months of the year.

As for the necessity of there being an overlap of territorial waters, it is assumed that the waters of the Northwest Passage were not internal on the basis of history before their enclosure by straight baselines. The assumption is based on the conclusion reached in Chapter 8 that the extension of Canada's territorial waters to 12 miles in 1970 resulted in an overlap of territorial waters in Barrow Strait. This means that, since that date, all of the routes of the Northwest Passage must cross the territorial waters of Canada. The effect of the 1970 extension was accurately expressed by the Secretary of State for External Affairs at the time of the introduction of the bill in the House of Commons when he stated:

> Since the 12-mile territorial sea is well established in international law, the effect of this Bill on the Northwest Passage is that under any sensible view of the law, Barrow Strait, as well as the Prince of Wales Strait, are subject to complete Canadian sovereignty.[57]

In light of the stated assumption, the application of the geographic criterion leads to the conclusion that the Northwest Passage constitutes a legal or territorial strait, in that it connects two parts of the high seas, or exclusive economic zones, and presents an overlap of territorial waters.

The *functional criterion*, as applied by the International Court in the *Corfu Channel Case* requires that a strait has been a useful route for international maritime traffic, as evidenced mainly by the number of ships using the strait and the number of flags represented, before it can be classified as an international strait. When this criterion is applied to the Northwest Passage, it becomes readily evident that it fails to be met since, in its 80-year history of exploratory navigation, the Passage has seen only 45 complete transits and, of these, 29 were by Canadian ships. The 16 foreign crossings comprised 11 American ships, 1 Norwegian, 1 Dutch, and 1 Japanese, 1 Bahamian and 1 Liberian. The historic Norwegian crossing of 1903–6 by Amundsen was one of discovery, the Dutch and Japanese crossings were for adventures, and the Bahamian and Liberian were pleasure cruises. Aside from the first discovery crossing, all others were proceeded by a request for and grant of permission.

As for the American transits, three of them were accomplished by a squadron of icebreakers in 1957, performing hydrographic surveys during

the joint Canadian–American establishment of the Distant Early Warning System, and all three ships were led through the narrow Bellot Strait
by H.M.C.S. *Labrador*. Two American submarine crossings took place
to test the feasibility of submerged transits of the Northwest Passage. The
U.S.S. *Seadragon* in 1960 had obtained Canada's permission and had a
Canadian representative aboard in the person of Commodore O. C. S.
Robertson, and the U.S.S. *Skate* in 1962 made its crossing within the
context of U.S.–Canada defence arrangements. Five of the other six
American transits were made in 1969, when the S/T *Manhattan* tanker
loaded with water made its feasibility voyage in Route 1 and was
accompanied for part of the voyage by the U.S. icebreakers *Staten Island*
and *Northwind*. The *Manhattan* had a Canadian representative aboard in
the person of Captain T. C. Pullen and was escorted by the Canadian
icebreaker *John A. Macdonald*. The last American crossing was by the
American Coast Guard icebreaker *Polar Sea* which re-deployed from
Thule, Greenland, to the Chukchi Sea in August 1985, to carry out
oceanographic work. The American Government refused to ask permission and, in spite of Canada's cooperation, the refusal resulted in the
decision to enclose the waters of the Northwest Passage by the establishment of straight baselines around the Arctic Archipelago.

It is clear from the above review that by no stretch of the imagination
could the Northwest Passage be classified as an international strait. Those
who maintain that the Passage may be so classified obviously confuse
actual use with *potential use*. The latter test is the one used by American
courts to determine whether a waterway is navigable or not. In such a
case, as explained by Professor Baxter, it is the capacity for navigation
which is the effective criterion.[58] This is not the criterion of actual use
required in international law and applied by the International Court in the
Corfu Channel Case. In addition, it must be pointed out that, with the
possible exception of the cruise ships *Lindblad Explorer* and *World
Explorer*, both of which asked permission, not one of the few foreign
transits could be characterized as constituting commercial navigation.

Was there a right of passage before the straight baselines?
Based on the conclusion that the straits of the Northwest Passage
consisted of territorial waters and high seas (or exclusive economic zone)
before the establishment of straight baselines, those waters had to be
subject to the right of innocent passage. This right applied as it did in any
territorial waters, in spite of the fact that the Northwest Passage did not
qualify as an international strait. The absence of such qualification,

however, meant that Canada could have suspended the innocent passage of foreign ships for security reasons.

The only question which remained was whether such right of innocent passage existed in favour of *all* ships, including warships. This has been a controversial question for a long time and, indeed, still is today to a certain extent. In the *Corfu Channel Case*, the International Court carefully avoided pronouncing on the question of innocent passage of warships through the territorial sea as such. It confined its judgment to the strict question at issue, namely, the right of innocent passage of warships in peacetime through international straits.

In preparing the draft of the 1958 Geneva Convention, the International Law Commission fully debated the question of passage of warships. It began by granting passage to warships without prior authorization or notification but, due to the strong opposition of certain members and the comments received from governments, it provided in its final draft article that prior authorization was necessary. It specified, however, that it should normally be granted.[59]

At the 1958 Conference, after much renewed debate, no specific provision on the right of innocent passage of warships was retained, but all provisions relating to innocent passage were made applicable to all ships.[60] The Conference also approved the special article, providing for the expulsion of warships from the territorial sea in case of refusal to comply with the regulations of the coastal State.[61]

As a result of the above, the right of innocent passage in favour of warships in territorial waters continued to be questioned after the Conference, and the Soviet Union attached a reservation to its ratification of the Territorial Sea Convention, which provided for prior authorization. The reservation is consonant with the interpretation put on the Convention by Professor Tunkin and other Soviet jurists.[62] On the other hand, the overwhelming view of international lawyers and commentators is that the Convention does grant the right of innocent passage to warships.[63]

On this question, the well-known legal historian, Professor Verzijl, made a thorough review of the proceedings and deliberations of the 1958 Conference, both in Committee and in Plenary, and came to the following conclusion:

> After the rejection of the exceptional provision for passing warships, their freedom of passage is clearly covered by the general rules in subsection A 'applicable to all ships'. No falling back upon a pretended contrary rule of customary law, said to allow a coastal State to prevent the passage of warships without its authorization – a rule, for the rest, resolutely opposed by

others – would seem to be of any judicial avail against the explicit provisions and the logical context of section 3 of part 1 of the first Convention.[64]

In spite of the controversy on the proper interpretation of the 1958 Territorial Sea Convention relating to the right of innocent passage of warships, the 1982 Convention followed exactly the same approach. In Part 2, on the territorial sea and contiguous zone, Section 3, entitled 'Innocent Passage in the Territorial Sea', begins with subsection, 'Rules Applicable to all Ships'. This subsection is of general application to *all* ships and makes no mention of warships in particular. However, as in 1958, the new Convention contains the same article permitting the expulsion of a warship, which does not comply with the regulations of the coastal State concerning passage through its territorial sea and disregards a request for compliance.[65]

As to the right of innocent passage on the part of submarines, both the 1958 and 1982 Conventions provide that they have an obligation to navigate on the surface and show their flag. In 1982 the text enlarges this obligation to 'other underwater vehicles'.[66] Considering that the innocent passage of submarines would normally be warships, it is an additional reason to conclude that the intention of the drafters of both the 1958 and the 1982 Conventions was to make the right of innocent passage applicable to warships.

Even the Soviet Union has come around to recognizing the right of innocent passage of warships through territorial waters. This should not be too surprising, considering the exceptional development of its naval forces, including the important Northern Fleet stationed off the Kola Peninsula. The first official sign of this change of position is found in the Draft Articles on the Territorial Sea introduced at the Third Law of the Sea Conference in 1974, in which the right of innocent passage was stated to 'apply to foreign warships'.[67] Then its 1982 'Law on the State Boundary of the U.S.S.R.,' specifically provided that 'foreign warships, and also submarine means of transport, shall effectuate innocent passage through the territorial waters (territorial sea) of the U.S.S.R. in accordance with the procedure established by the U.S.S.R. Council of Ministers'.[68]

This new legislation having come into force on March 1, 1983, the Council of Ministers adopted a Decree on April 28 of the same year, specifying that 'foreign warships enjoy the right of innocent passage through the territorial waters (territorial sea) of the U.S.S.R. . . . .'.[69] Neither the Law nor the Decree provides for any prior authorization. Such authorization is required for warships only when putting in internal waters.[70] The conditions for the exercise of innocent passage are basically the same as those provided for in the 1982 Law of the Sea Convention[71] and submarines are required to navigate on the surface.[72]

In spite of the above analysis which points to a general acceptance of the right of innocent passage in favour of all ships, it must be stated that such right is still objected to by certain States. At the last session of the Law of the Sea Conference in 1982, some 20 states (mostly developing coastal States) pressed for an amendment to the Draft Convention which would have required either prior authorization or notification for the innocent passage of warships. The proposed amendment was eventually withdrawn on condition that the Conference President read a statement in plenary that the withdrawal was without prejudice to the views of States requiring prior authorization or notification. In conformity with their objection, some nine States attached an interpretative declaration to their signature of the 1982 Convention, reserving their position on this question.[73]

Although Canada has consistently expressed a restrictive view of the right of innocent passage with respect to tankers, it has not taken any comparable position in relation to warships. It has no legislation on the question and no special rule in its Notices to Mariners. It is the understanding of this writer that Canada relies on customary international law on the matter.

The above review leads to the conclusion that the right of innocent passage of all ships in the territorial sea has been accepted by the great majority of States, including coastal States, and forms part of customary international law. Accordingly, such a right existed in the Northwest Passage before its enclosure by straight baselines in 1985.

Is there a right of passage since the straight baselines?

Under customary international law, as applied by the International Court in the *Fisheries Case* of 1951, there is no right of passage in waters enclosed by straight baselines. This is the result, regardless of the previous status of the newly enclosed waters. However, the straight baseline system was modified in this respect when incorporated in the 1958 Territorial Sea Convention. The latter made the enclosed waters subject to the right of innocent passage, if they were previously territorial waters or high seas. Since Canada is not a Party to the Convention, it would naturally rely on customary law for the validity of its straight baselines and the resulting legal status of internal waters, if that status cannot rest on an historic title. Such reliance might not be completely secure, however, when one considers the possibility of the 1958 Convention provision having become part of customary law.

The 1958 Convention came into force in 1964, with 22 ratifications. Since then, the number of ratifications, accessions and successions has raised the membership to 45. These 45 States, however, include only 21 of the 60 States which have used the straight baseline system and only two of the additional 12 which have adopted enabling legislation (see Table 1). In other words, there are 49 States that have actually used straight baselines or have adopted enabling legislation but have not become Parties to the Convention.

In the circumstances, it seems impossible to conclude that the acceptance of the 1958 provision for innocent passage in newly enclosed internal waters has been so general as to become legally binding on all States. The International Court made it clear, in the *Gulf of Maine Case* of 1984, that more was required before reaching an affirmative conclusion as to the existence of customary law. It held, in that case, that the equidistance method of continental shelf delimitation had not become a rule of customary law and neither had it been adopted into such law as a method to be given preference over others.[74] The Court so held in spite of the fact that the 1958 Continental Shelf Convention, which provided for such rule, had been in force since 1964 and some 54 States were Parties to it. In addition, the great majority of at least 80 delimitation agreements, concluded since 1958, had been based on the equidistance method, either strict or modified. Consequently, it is highly unlikely that the Court would hold that Article 5 of the Territorial Sea Convention has become binding on all States on the basis of a newly created customary law. This is all the more unlikely that the provision represented an important departure from existing customary law.

Accordingly, the conclusion is that no right of innocent passage exists in the new internal waters of the Northwest Passage. And, as properly remarked by Professor O'Connell, 'the fact that there is access to the sea at either end of landlocked waters does not deprive these of the essential characteristics of internal waters'.[75] Conversely, the status of internal waters does not affect the nature of the Northwest Passage as a strait (in fact, a series of straits) and perhaps capable of being internationalized.

14.3. Internationalization of the Northwest Passage

Three basic questions must be answered: (1) Could the Northwest Passage become an international strait; (2) What right of passage would apply in that event; and (3) What measures could Canada take to prevent the internationalization?

Could the Northwest Passage be internationalized?

The possible internationalization of the Northwest Passage will depend on the degree of international shipping which will take place and the measures which Canada will take to exercise control of such shipping.

On the question of international shipping, it has already begun in the eastern part of the Passage to transport minerals from Nanisivik Mine, south of Lancaster Sound, and Polaris Mine, north of Barrow Strait. Also, it seems to be only a question of time before regular shipping takes place from Melville Island, north of Viscount Melville Sound, and from the Beaufort Sea along the full length of the Northwest Passage. Oil will probably come from both the Canadian and the Alaskan sides of the Beaufort Sea. An estimate, as to a realistic time-frame for the latter shipping to begin, would place it between 1995 and the year 2000.

Although the threshold use in the *Corfu Channel Case* of 1951 was fairly high, a considerably lower threshold would probably suffice in the Northwest Passage. Because of special factors such as the remoteness of the region, the difficulties of navigation and the absence of alternative routes, comparatively little use might be required. A pattern of international shipping across the Passage, developed over relatively few years, might be held sufficient to make it international. It has already been recognized by the Permanent Court of International Justice that the application of general principles of law to the Arctic regions must take into account the special local conditions.[76]

Of course, an internationalization of the Passage presumes that Canada would allow the passage of foreign ships, without taking appropriate measures to insure effective control over such ships and the waters in question. It must be remembered that, although the enclosure of those waters has resulted in a sovereignty for Canada which is as complete as over the islands, such sovereignty must be maintained and this can only be done by the exercise of effective control. This means that practical measures must be taken. These will be discussed after an examination of the right of passage which would apply in an internationalized Northwest Passage.

What right of passage would apply after internationalization?

On the insistence of the major Maritime Powers and as part of an integral package, a consensus was finally reached at the Third Law of the Sea Conference as to the type of passage applicable in straits used for international navigation. This new right is called 'transit passage' and is defined in the 1982 Convention as follows:

Transit passage means the exercise in accordance with this Part of the freedom of navigation and overflight solely for the purpose of continuous and expeditious transit of the strait between one part of the high seas or an exclusive economic zone and another part of the high seas or an exclusive economic zone.[77]

Since the expression 'transit passage' replaced that of 'free passage', which the United States and the Soviet Union had in their proposals,[78] the question arises as to whether this new right of passage is more restrictive than the one which the two great Powers wanted. More specifically, what are the limitations of 'the freedom of navigation and overflight' in international straits as distinguished from the same freedoms on the high seas? The definition contains two limitations: first, the freedom has to be exercised 'in accordance with this Part', and second, it must be 'solely for the purpose of continuous and expeditious transit'. These limitations are expressed in the form of duties imposed on ships and aircraft during transit, and are found in Articles 39, 40 and 41.

Article 39 specifies certain duties for ships and aircraft, either together or separately. Both ships and aircraft must proceed without delay, refrain from any threat or use of force, and refrain from any activity not incidental to their normal modes of transit unless rendered necessary by *force majeure* or distress. In addition, there is a duty imposed on ships to comply with the generally accepted international regulations relating to safety at sea and pollution prevention and control.

Article 40 prohibits ships from carrying out research and survey activities, without the prior authorization of the bordering States, and Article 41 requires them to respect applicable sea lanes and traffic separation schemes.

Such is the extent of the limitations imposed by the Convention on the freedom of navigation through international straits. Unlike the case of the exercise of the right of innocent passage, there is no list of prohibited activities applicable to the exercise of the right of transit passage. In these circumstances, one must conclude that the removal of the words 'the same freedom of navigation as on the high seas', qualifying the type of the passage applicable to international straits, and their replacement by the qualifying words just discussed, did not change the essential nature of the right originally contemplated.

As in the case of innocent passage through the territorial sea, the question arises here as to whether the right of transit passage in international straits applies to warships, since there is no mention of them in the Convention. The absence here is even more complete in that there is no provision whatever relating to submarines. However, a careful reading of the relevant provisions of the Convention in light of the numerous statements made by the Maritime Powers at the time of the Conference, makes it abundantly clear that the intention was to include warships.

In the first place, the Convention stipulates that '*all* ships and aircraft enjoy the right of transit passage, which shall not be impeded'. The applicability of this right to all ships is reinforced by a subsequent provision limiting the laws and regulations which may be adopted by bordering States. This limitation specifies that 'such laws and regulations shall not discriminate in form or in fact among foreign ships'.[79] One cannot conclude from this alone, however, that it is clear and unambiguous that the right of free transit includes that of submarines.

It is submitted that the ambiguity or obscurity of the provisions already referred to is clarified in that part of Article 39 which specifies the common duties of ships and aircraft while exercising their right of transit passage. That Article provides that they shall 'refrain from any activities other than those incidental to their *normal modes* of continuous and expeditious transit unless rendered necessary by *force majeure* or by distress'.[80]

As succinctly pointed out by Professor O'Connell, since submarines are by definition underwater vehicles, submerged passage is a "normal mode"' of operation for such craft'.[81] And, as further explained by Professor O'Connell, the reason for the express prohibition against submerged passage in the territorial sea and the implied permission for submerged transit in international straits 'rests on the essentially different juridical character of the territorial sea from straits'.[82]

The above interpretation, based solely on the terms of the Convention itself and the ordinary meaning of those terms in their context, is fully confirmed by an examination of the circumstances of the conclusion of the Convention. True, the preparatory work is not very satisfactory in that it does not include a complete record of the proceedings of the Conference; however, the circumstances of the conclusion of the Convention, as they appeared throughout the Conference, unquestionably confirm the textual interpretation given above.

Professor Moore is in a good position to summarize those particular circumstances, since he served for a while as the United States ambassador to the Conference and, in that capacity, headed the U.S. team that participated in the development of the navigation and security provisions. Referring to the circumstances prevailing at the Conference on this question, Professor Moore concludes:

This evidence includes repeated statements by the United States, the Soviet Union, and other maritime States concerning the essentiality of transit passage (including submerged transit and other incidents), followed by absence of objection to the I.C.N.T. text by these parties; efforts by extreme strait States to reopen the text and prohibit submerged transit or overflight of straits; common and uncontradicted use of special terms during negotiations such as 'freedom of navigation' and 'in the normal mode' to include submerged transit; and on-the-record exchange indicating that the United Kingdom's transit passage text contemplated overflight and submerged transit.[83]

Indeed, the major Maritime Powers, in particular the United States, repeatedly stated throughout the Conference that freedom of naval mobility through international straits, in virtually the same way as on the high seas, was absolutely essential to their national security. The unswerving position of the United States was that it could not agree to an extension of a 12-mile territorial sea doing away with a strip of high seas from some 113 straits (according to its own estimate), unless it retained the freedom of navigation and overflight through and over such straits for its ships and aircraft.

This condition was emphasized not only by the United States delegation at the Conference, but also by the Secretary of State outside the Conference. In a prepared speech delivered to the Annual Meeting of the American Bar Association in August 1975, Mr Kissinger stated:

> After years of dispute and contradictory international practice, the Law of the Sea Conference is approaching a consensus on the twelve-mile territorial limit. We are prepared to accept this solution, provided that the unimpeded transit rights through and over straits used for international navigation are guaranteed.[84]

The Secretary of State concluded on this question by saying, 'we will not join in any agreement which leaves any uncertainty about the right to use world communication routes without interference'.[85]

The United States having decided to stay outside of the Convention, the interesting question now remains whether it could exercise this new right of passage after the Convention comes into force. True enough, a 12-mile territorial sea has now unquestionably become part of customary international law, as well as the right of non-suspendable innocent passage in international straits, but this is not the case for the new right of transit passage. The most that can be said for the latter is that it might be at the beginning of an emerging state of customary law. This means that, if the Convention should come into force before the complete emergence of this new right, the United States, as a Non-Party to the Convention, could not exercise the right of transit passage in the Northwest Passage even if it were internationalized.[86]

It must be added that the above conclusion does not seem to coincide with the position of the United States. The conclusion of the Department of Defence representative and Vice-Chairman of the U.S. delegation at the Law of the Sea Conference, expressed in 1983, is that 'the basic parameters of the Convention's transit passage regime, including the rights of submerged transit and of overflight, are in fact reflective of rights available to all nations who sail the world's oceans'.[87] Indeed, it has been stated that the United States never conceded that, under customary international law, 'the regime of innocent passage applied in straits, at least those straits wider than 6 miles'.[88] In other words, the freedom of navigation in such straits, even if not used for international navigation, would be greater than under the regime of innocent passage. This was confirmed by an official of the State Department, prior to the crossing of the *Polar Sea* in June 1985, who stated with reference to the Northwest Passage that 'in the territorial sea all vessels have a right of innocent passage and in a strait to a transit passage regime'.[89] Such statements would indicate that the United States considers the right of transit passage to be applicable to all straits, including the Northwest Passage. Consequently, it is not surprising that Canada judged it necessary to enclose the Passage by straight baselines and take measures to prevent the internationalization of the Passage.

Measures to prevent the internationalization of the Passage

In his statement of September 10, 1985 to the House of Commons, the Secretary of State for External Affairs made it clear that it was the policy of the Government 'to make the Northwest Passage a reality for Canadian and foreign shipping, as a Canadian waterway'.[90] This policy is basically the same as that announced by the Trudeau government in 1969, at the time of the *Manhattan* voyage, when the Prime Minister himself stated that 'to close off those waters and to deny passage to all foreign vessels in the name of Canadian sovereignty . . . would be as senseless as placing barriers across the entrance of Halifax and Vancouver harbours'.[91] The 1969 policy statement was followed the next year by special legislation to protect the Arctic marine environment and the 1985 statement was accompanied by the actual enclosure of the Northwest Passage with straight baselines. In addition, in 1985, special measures were announced to eventually gain effective control over the waters of the Passage and thus prevent its internationalization. Such measures are also designed to protect Canada's most important interests in the Arctic: the special marine environment, the welfare of the Inuit population and the national security of the country. The purpose here is to comment briefly on the measures which were already in force and those which were announced – aside from the establishment of the straight baselines which were dealt with earlier – and suggest certain additional measures which could be taken.

Measures already in force
At least three measures had already been taken by Canada, prior to the 1985 *Polar Sea* incident, to exercise a certain degree of control of the Northwest Passage.

Pollution Prevention Legislation (1970) After the voyage of the *S.S. Manhattan*, Canada adopted the Arctic Waters Pollution Prevention Act and made it applicable to all the waters of its Arctic Archipelago north of the 60th parallel, as well as to the waters outside the Archipelago up to 100 nautical miles.[92] The Act provides for absolute liability on the part of both ship and cargo owners for the deposit of waste.[93] The Act also provides for the establishment of shipping safety control zones and the prescription of certain standards of construction, navigational aides, manning, pilotage and icebreaker assistance.[94] Foreign public ships may be exempted from such prescription if they meet standards substantially equivalent.[95] Pollution prevention officers may, in certain circumstances, board a ship and even seize such ship and its cargo.[96] So far, it has not been necessary to exercise such powers.

Pollution Prevention Regulations and Control Zones (1972) Pursuant to the 1970 legislation, the following were adopted in 1972: the Arctic Waters Pollution Prevention Regulations, the Arctic Shipping Pollution Prevention Regulations and the Shipping Safety Control Zones Order. These regulations and the Order insure that ships entering Arctic waters meet the pollution prevention requirements, these varying with the zones to be traversed. The Arctic waters, as defined in the 1970 legislation, are divided into 16 zones (see Figure 18), requiring a particular class of ship depending on the time of year (see Table 2). For instance, a Class 10 icebreaker (or ice-strengthened tanker) would be required to navigate in M'Clure Strait on a year-round basis, whereas a Class 8 would suffice from July to mid-October.

Voluntary Traffic System NORDREG (1977) Introduced in July 1977, this system encompasses both the Eastern and Western Arctic waters within a single vessel traffic management system.[97] The main purpose is to enhance the safety of navigation and prevent pollution. The system applies to all ships of 300 tons or over, but efforts are made to obtain information on all vessels entering Arctic waters. Normally, a clearance request should be made 24 hours before entering the NORDREG Zone and, once underway within the Zone, reports should be made daily to the Coast Guard Traffic Centre at Frobisher Bay. The role of the Centre is to exercise control over Arctic navigation and to provide ships with navigational information and services. So far, the system has not been made mandatory and the word 'voluntary' appears at the beginning of the Notice to Mariners.

It should be noted that none of the above measures constituted an assertion of sovereignty on the part of Canada. They represented simply an exercise of jurisdiction for one basic purpose: pollution prevention and control. Nonetheless, when they were adopted or provided for by Canada in 1970, there was some doubt in international law as to the power of a coastal State to exercise such jurisdiction in maritime zones which included areas of the high seas at that time.

The United States strongly objected to Canada's claim of jurisdiction, prompting Canada to embark upon intense diplomatic efforts in a number of international fora to obtain recognition of international law principles which would serve as bases for its legislation. These efforts were pursued mainly at three international conferences: the Stockholm Conference on the Human Environment of 1972, the London Conference of the International Maritime Consultative Organization on the Prevention of Pollution by Ships in 1973, and the Third United Nations Law of the Sea Conference which began in 1974.

At the 1975 session of the Law of the Sea Conference, a form of special Arctic clause was inserted in the first Negotiating Text. It provided that coastal States could adopt special protective measures in areas within their exclusive economic zone, where exceptional hazards to navigation prevailed and marine pollution could cause irreversible disturbance of the ecological balance. In 1976 the provision was enlarged to enable coastal States themselves to enforce such protective measures, instead of leaving the enforcement to the flag State.

The above provision was retained without change in the subsequent Negotiating Texts of 1977, 1979 and 1980, as well as in the Convention itself signed in December 1982. The special Arctic clause in question reads as follows:

Coastal States have the right to adopt and enforce non-discriminatory laws and regulations for the prevention, reduction and control of marine pollution from vessels in ice-covered areas within the limits of the exclusive economic zone, where particularly severe climatic conditions and the presence of ice covering such areas for most of the year create obstructions or exceptional hazards to navigation, and pollution of the marine environment would cause major harm to or irreversible disturbance of the ecological balance. Such laws and regulations shall have due regard to navigation and the protection and preservation of the marine environment based on the best available scientific evidence.[98]

There are at least five elements in this special clause which should be highlighted. First, the coastal State not only has the right to adopt special protective measures, but it may also enforce them against foreign ships. Second, the permitted regulations may be of a preventive nature and may, therefore, apply to the design, construction, manning and equipment of foreign ships. Third, this right applies within the limits of the 200-mile exclusive economic zone and, *a fortiori*, to the limits of the territorial sea. Fourth, such measures may only be taken in areas where ice is present most of the year, where it creates exceptional hazards to navigation, and where pollution could cause major harm to, or irreversible disturbance of, the ecological balance. Fifth, the regulations of the coastal State must have due regard to the imperatives of navigation and the best scientific data available relating to the protection of the marine environment.

It should be noted that the first two elements just mentioned above circumscribe the extent of the right conferred on the coastal State, whereas the other three define the physical requirements for the valid exercise of that right. In addition, it is important to realize that the exercise of that right by the coastal State is not subject to any control whatever by the International Maritime Organization.

The special Arctic clause removes any doubt as to the international validity of Canada's 1970 legislation and, having regard to the wide consensus which the clause has received, particularly on the part of other Arctic States, it may now be regarded as accepted in customary international law.[99] This special power, however, is limited to situations where there is a threat to the marine environment. It covers basically the case of tankers carrying oil or gas. It is doubtful that the special clause could be invoked to interfere with the passage of foreign icebreakers, all the more so if the Northwest Passage were to be internationalized. In that event, and even assuming that the special clause would continue to apply,[100] Canada would have little if any control over such icebreakers, particularly if they are armed and may be classified as warships. Indeed, under the new right of transit passage which would apply, the passage of warships could not be interfered with. This unimpeded freedom of navigation applies also to submarines which can transit in their normal mode of navigation, that is, submerged. In the Arctic, this means navigation under the ice cover, making detection virtually impossible.

Measures announced in September 1985
In its September 1985 policy statement, the Government announced a number of measures, four of which will be examined here. Those measures relate to: icebreaking services, surveillance overflights, a naval presence in the Eastern Arctic and a user State agreement with the United States.

Class 8 Icebreaker The Government has announced that it has 'decided to construct a Polar Class 8 icebreaker'.[101] This is the logical next step to the one taken by the previous Government in early 1983 to approve funds for the design, but not construction, of such an icebreaker. The icebreaker, capable of breaking ice 8 feet thick, at a continuous speed of 3 knots, will enable Canada to exercise a reasonable degree of control over surface navigation in the Northwest Passage, at least in its Route 1 (see Figure 19) through Prince of Wales Strait. The icebreaker should be so designed as to permit the installation of weapons systems and thus ensure a quasi-military presence in the area.[102] If the Coast Guard is to assist the Canadian Navy in performing a limited sea-denial role, however, there should be greater coordination of activities between the two.[103]

In addition, the Class 8 icebreaker would perform such tasks as escorting cargo ships, enforcing pollution prevention laws and regulations, pursuing oceanographic research and responding to calls for search and rescue. This it would be able to accomplish on a year-round basis in any part of Route 1, insure a certain presence and thus strengthen Canada's assertion of sovereignty over those waters. Of course, a single icebreaker will not be sufficient for surveillance and control of year-round shipping, but it will probably suffice until such shipping begins. The question which remains is what will Canada do during the six- or seven-year-period of construction of the Class 8. It is strongly suggested that efforts be made to make contingency arrangements for the leasing of a Class 8 (or equivalent) icebreaker. This will pose a practical problem of considerable magnitude, since the availability of such icebreakers (either already built or ready to be fitted) is very restricted, but it should not prove to be beyond the imagination and determination of Canadians.

Surveillance overflights Up to September 1985, only about 16 sets of reconnaissance overflights missions, each about three or four days long, were undertaken each year.[104] The Government has now decided to take 'immediate steps to increase surveillance overflights of our Arctic waters'.[105] It is not known at the time of writing the extent of the intended increase, neither as to the number of overflights nor as to the area to be covered. These measures, of course, are separate and apart from those agreed upon with the United States in March 1985 to modernize the North American Air Defence System, although this modernization will include the upgrading of some existing northern airfields to permit their use by fighter aircraft.[106] Whatever increase takes place in surveillance over-flights, these should be appropriately coordinated with naval activities.

Naval presence in Eastern Arctic Announced also in September 1985 were 'plans for naval activity in eastern Arctic water in 1986'. So far, there is virtually no naval activity north of the Labrador coast. In its 1983 report, the Sub-Committee on National Defence of the Senate Committee on Foreign Relations found it baffling that Maritime Command had not been provided with additional equipment to exercise appropriate surveillance in spite of the enormous additions to Canada's maritime jurisdictional claims since publication of 'Defence in the 70s'.[107] The Sub-Committee added that 'there is a requirement for Canada's maritime forces to be equipped to perform a sea-denial role in waters over which Canada claims jurisdiction'.[108] Canada having now confirmed its sovereignty claim over the waters of the Arctic Archipelago by the establishment of straight baselines, it had no choice but to provide for certain naval activities to protect those waters. What these activities will be is not known, but it is obvious that they should insure an adequate surveillance of both surface and sub-surface areas at the main entrances of the Northwest Passage. Sub-surface surveillance requires special attention and certain measures will be suggested later.

User State Agreement with U.S.A. The Policy Statement announced also 'immediate talks with the United States on cooperation in Arctic waters, on the basis of full respect for Canadian sovereignty'.[109] Presumably, this must envisage the negotiation of what could be called a 'User State Agreement', whereby the United States would expressly recognize

Canada's sovereignty over the enclosed waters of the Arctic Archipelago, including those of the Northwest Passage, and Canada would guarantee the United States the right for its merchant ships to use the Passage under stipulated conditions.[110] The Agreement would not grant a servitude in perpetuity but would be in force for a fixed term, of let us say ten years, and would then continue into force subject to termination on a year's notice. This is, for example, what was recently provided for in the Agreement for the Modernization of the North American Air Defence System.[111] The passage of warships and submarines would be permitted only by virtue of special authorization or under specific provisions of continental defence arrangements. The stipulated conditions relating to merchant vessels would also provide for the compulsory use of icebreaking, pilotage and other navigational services of Canada. Similar agreements should be concluded with other States interested in using the Northwest Passage to insure Canada's continuing effective control.

Additional measures suggested
In addition to the above measures, the Government stated that it would announce further steps 'to provide more extensive marine support service'.[112] Those additional steps have not yet been announced but, presumably, they will relate to matters such as the following: compulsory vessel traffic system; eventual upgrading icebreaking services; compulsory pilotage; other navigational aids and services and under-ice submarine detection.

Compulsory vessel traffic system Now that Canada's sovereignty over the waters of the Arctic Archipelago has been confirmed, the existing traffic system NORDREG should be made compulsory for all vessels entering those waters. As appropriately suggested in the Final Report of the Environmental Assessment Panel for the Beaufort Sea in 1983, 'if a vessel is disabled or sinking in Arctic waters, or has encountered some other emergency, government's responsibility must be absolute and its actions must be swift and unencumbered by jurisdictional or communications problems'.[113] Consequently, the Beaufort Sea Panel recommended that NORDREG be made mandatory. It is now possible to implement that recommendation and this should be done without delay.[114]

Class 10 icebreaker For the short term, the Class 8 icebreaker which has been authorized should meet Canada's requirements reasonably well. However, as admitted by the Commissioner of the Coast Guard back in 1982, one such vessel would not be sufficient for year-round surveillance and adequate control of foreign shipping.[115] In addition, a Class 10 is required by Canada's own regulations for year-round navigation in M'Clure Strait. In these circumstances, the Senate Sub-Committee on National Defence recommended in 1983 that Canada give consideration to building a class 10 icebreaker.[116] In comparison, it is interesting to note that, in the Northeast Passage, the Soviet Union had a fleet of at least 22 icebreakers by 1983, of 10,000 shaft horse power or more, three of them being of the Arktika class with 75,000 s.h.p.[117] When the first Arktika class icebreaker was built in 1977, it made a trip to the North Pole in August of that year and encountered no problem. When hydrocarbon resources are transported from the Beaufort Sea to the American eastern seaboard, or perhaps from Newfoundland's offshore to Japan, Canada should be ready with the necessary year-round icebreaking support services. Otherwise, foreign shippers will provide such services themselves. Japanese firms are already preparing for the building of ice-breaking tankers.[118]

Compulsory pilotage Because of the rather unpredictable ice conditions in Arctic waters, it might be required of foreign shippers to retain the services of specially trained pilots in certain areas. The waters of the Arctic Archipelago could be divided into pilotage districts or zones. If this measure should prove unnecessary, perhaps the services of ice advisers could be made compulsory. And, for this task, members of the local Inuit population could probably become most effective. In this way, the protection of the marine environment would be more secure and so could Canada's control over those waters.

Navigational aids and services generally In its position statement of 1982 to the Beaufort Sea Environment Assessment Panel, the Department of Transport recognized that it should eventually be in a position to provide the following services: marine navigational aids, icebreaking and escorting, marine search and rescue, marine emergencies/pollution control, marine mobile communications services, ports, harbours and terminals, vessel inspection services, vessel traffic management, marine re-supply administration and support, pilotage and training.[119] The position paper specified that these services would be in addition to those such as hydrography, oceanography, ice properties, ice distribution and movements, meteorology, dredging implementation, and customs services which other government departments would continue to provide. The sooner Canada can insure these various services to foreign shippers, the sooner it will insure effective control over its Arctic waters.

Submarine detection capability It was expressly admitted in the 1985 Arctic sovereignty statement that 'Soviet submarines are being deployed under the Arctic ice pack, and the United States Navy in turn has identified a need to gain Arctic operational experience to counter new Soviet deployments'.[120] This possibility had been implied back in 1971 when the *White Paper on Defence* underlined that Canada had 'only very limited capability to detect submarine activity in the Arctic'.[121] The White Paper added that 'it might be desirable in the future to raise the level of capability so as to have subsurface perimeter surveillance, particularly to cover the channels connecting the Arctic Ocean to Baffin Bay and Baffin Bay to the Atlantic'.[122] Consequently, the Government announced that it was 'undertaking research to determine the costs and feasibility of a limited subsurface system to give warning of any unusual maritime activity'.[123]

The announcement does not seem to have been followed by the adoption of satisfactory measures, since the Senate Sub-Committee on National Defence found in 1983 that there was still no adequate surveillance of the Northwest Passage for submarine activities. The Sub-Committee concluded that such surveillance 'could be provided, for the time being, by conventionally powered submarines stationed at the entrance and the exit of the Passage',[124] and that installation of an economically and technically feasible bottom-based sonar system might also be desirable.[125] It added that 'should the frequency of nuclear submarine transits through Arctic waters rise substantially, Canada might have to contemplate obtaining nuclear submarines of its own', noting that asking its allies to patrol the area would hold real dangers since the United States did not recognize Canada's claim over the Arctic waters.[126]

There are indications that an increase of submarine activities in Arctic waters might have already taken place. Admiral James D. Walkins, the American chief of naval operations, was reported to have disclosed in May 1983 that the Soviet Union had increased its submarine operations in the Arctic and that the United States had responded by beginning extensive submarine training under the ice pack.[127] Considering the possibility and advantages of using certain channels in the Canadian Arctic Archipelago as transit routes for attack and cruise-missile submarines,[128] Canada might soon find its Arctic waters being used as a training ground.

At the moment, Canada's anti-submarine warfare capability is still limited to open water and it has no submarine detection capability of its own for ice-covered areas. The least Canada must do, to protect its newly proclaimed sovereignty and its national security, is to develop an adequate submarine detection capability in the Arctic. Surely, it has a right to know what goes on in its own waters; indeed, it has a duty to find out. One way would be to install hydrophone or sonar barriers at appropriate intervals, on the seabed, in key passages of the Archipelago.[129] There might well be more effective means of detection and Canada should try to develop the best possible monitoring techniques and devices.

Whether Canada should go further and acquire means of deterrence, such as nuclear attack or killer submarines, is another question. The cost might be considered prohibitive and, in addition, this ultimate and direct mode of deterrence is probably more appropriately provided for in joint defence arrangements with the United States. In the circumstances, and for the moment, Canada should concentrate its efforts in the development of the most effective submarine detection capability for the ice-covered areas of its Arctic waters in general and those of the Northwest Passage in particular.

Notes to Part 4

1. For a study of the Northwest Passage before its enclosure by straight baselines, see Donat Pharand, *The Northwest Passage: Arctic Straits*, Martinus Nijhoff Publishers, 236 pages (1984).
2. Canadian Hydrographic Service, *Sailing Directions, Arctic Canada*, Vol. 1, at 1 (1982).
3. For a description of the ice regime in the eastern approaches, *ibid.*, at 156–9.
4. For a description of the various routes of the Northwest Passage, *ibid.*, at 159–63.
5. *Ibid.*, Vol. 3, at 199.
6. *Ibid.*, at 164.
7. *Ibid.*, at 126.
8. *Ibid.*, at 167.
9. *Ibid.*, Vol. 2, at 105.
10. *Ibid.*, at 76.
11. *Ibid.*, at 90.
12. The main sources used for this section are: *Sailing Directions, Arctic Canada*, Vol. 1 (1982); Andrew Taylor, *Geographical Discovery and Exploration of the Queen Elizabeth Islands* (1964); and Alan Cooke and Clive Holland, *The Exploration of Northern Canada 500 to 1920* (1978).
13. This refers to the crossing of the U.S. Coast Guard *Polar Sea*, in August 1985, which refused to ask Canada's authorization and thus prompted the Government's decision to establish straight baselines around its Arctic Archipelago.
14. The main sources used for the resource potential of the Alaska North Slope/Beaufort Sea area are the *International Petroleum Encyclopedia* (1983) and the *U.S. Geological Survey Circular 860* (1982).
15. *Maritime Services to Support Polar Resource Development*, at 46 (1981).
16. The main sources used are the *Canadian Energy Supply and Demand 1980–2000* (1981) and F. Bregha, *Canadian Arctic Marine Energy Projects*, Canadian Arctic Resources Committee, 52 pages (1982).
17. See *Canada's Energy Frontiers*, 19 pages, tabled in the House of Commons by the Minister of Energy, Mines and Resources on 30 October 1985, at 4.
18. *Beaufort Sea/Mackenzie Delta Environmental Impact Statement*, Vol. 2 (1982).
19. Senate of Canada, *Marching to the Beat of the Same Drum*, Transportation of Petroleum and Natural Gas North of 600, at 31 (March 1983).
20. The main source is the *International Petroleum Encyclopedia* (1983).
21. See Canadian Press report 'Historic Oil Shipment from Arctic Arrives', *Globe*

and Mail, at B10 (12 Sept. 1985). See also Indian and Northern Affairs, *Intercom*, Nov. 1985, at 9–11.
22. The main source is the *Beaufort Sea/Mackenzie Delta Environmental Impact Statement*, Vol. 5 (1982).
23. Indian and Northern Affairs, *Intercom*, July 1985, at 3.
24. See Indian and Northern Affairs, *Communique* No. 3–8205, at 7–12 (17 June 1982).
25. See Indian and Northern Affairs, *1982–1983 Annual Report*, at 12 (1983).
26. See *supra* note 24, at 6.
27. *Supra* note 19, at 46.
28. Transport Canada, *Position Statement to the Beaufort Sea Environmental Assessment Panel*, Appendix 1, at 1 (1982).
29. *Ibid.*, at 36.
30. Canada, *Proceedings of Standing Committee on External Affairs and National Defence*, No. 24, at 6 (22 May 1975); emphasis added.
31. For a review of those attempts, see D. P. O'Connell, *The International Law of the Sea*, Vol. I, at 309–16 (1982).
32. 1982 Convention, Art. 36.
33. See R. R. Baxter, *The Law of International Waterways*, at 4–5 and footnote 12 (1964) and Charles De Visscher, *Problems de confins en droit international public*, at 141 (1969).
34. E. Bruël, *International Straits*, Vol. 1, at 42–3 (1947).
35. *Ibid.*, at 43; emphasis added.
36. On these qualifying expressions, see also O'Connell, *supra* note 31, at 301–6.
37. (1949) I.C.J. Rep., at 29.
38. (1949) I.C.J. Pleadings, Vol. II, at 242.
39. *Supra* note 37.
40. *Supra* note 38, at 354.
41. *Supra* note 37, at 28.
42. *Ibid.*
43. *Ibid.*
44. Art. 17, *I.L.C. Yearbook*, Vol. II, at 258 (1956); emphasis added.
45. G. C. Fitzmaurice, 'The Law and Procedure of the International Court of Justice; General Principles and Substantive Law', 27 *B.Y.I.L.* 1–41, at 28 (1950); emphasis added.
46. *Supra* note 37.
47. Baxter, *supra* note 33, at 3; emphasis added.
48. *Ibid.*, at 9.
49. D. P. O'Connell, *International Law*, Vol. I, at 497 (1970); emphasis added.
50. D. P. O'Connell, *The International Law of the Sea*, Vol. I, at 308 (1982); emphasis added.
51. *Ibid.*, emphasis added.
52. 1982 Convention, Art. 36.
53. *Ibid.*
54. *Ibid.*
55. *Ibid.*, Art. 38, para. 1 and Art. 45, para. 1(a).
56. *Ibid.*, Art. 45, para. 1(b).
57. *Can. H. C. Debates*, at 6015 (15 April 1970).
58. *Supra* note 33, at 3, footnote 5.
59. See Commentary to Article 24, *I.L.C. Yearbook*, Vol. II, at 276–7 (1956).
60. See 1958 Convention, Articles 14 to 17.
61. *Ibid.*, Art. 23.

62. See G. Tunkin, 'The Geneva Conference on the Law of the Sea', 1 *Int'l Affairs* 47–52, at 49 (1958).
63. See in particular: P. Jessup, 'The U.N. Conference on the Law of the Sea', 59 *Colum. L. R.* 234–268, at 248 (1959); J. H. W. Verzijl, 'The United Nations Conference on the Law of the Sea, Geneva, 1958', 6 *Netherlands Int'l L. R.* 1–42, at 34 (1959), and R. R. Baxter, *supra* note 33, at 167–8.
64. J. H. W. Verzijl, *International Law in Historical Perspective*, Vol. IV, at 188 (1971).
65. 1982 Convention, Art. 30.
66. *Ibid.*, Art. 20.
67. See U.N. Doc. A/CONF. 62/C.2/L.26.
68. Text of the Law translated by Wm. E. Butler and reproduced in 22 *International Legal Materials* 1055–1067, at 1060; Art. 13. See also *Law of the Sea Bulletin*, No. 4, at (Feb. 1985).
69. Decree reproduced in 24 *Notices to Mariners* (in Russian) 42–7 (1983) and translated by E. Franckx; Art. 8.
70. *Supra* note 68, Art. 14.
71. *Supra* note 69, Arts. 11 and 12.
72. *Supra* note 68.
73. See *Law of the Sea Bulletin*, No. 1, at 17–18 (Sept. 1983).
74. See *Delimitation of the Maritime Boundary on the Gulf of Maine Area (Canada v. United States of America)*, para. 107, 12 Oct. 1984.
75. *Supra* note 50, at 385–6.
76. See generally the *Eastern Greenland Case*, Denmark v. Norway, (1933) P.C.I.J. Rep., Ser.A/B, No. 53.
77. 1982 Convention, Art. 38, para. 2.
78. See 1971 Draft Articles of the United States, A/AC.138/S.C.II/L.4 and the 1972 Draft Articles of the Soviet Union, A/AC.138/S.C.II/L.7.
79. 1982 Convention, Art. 38, para. 1.
80. *Ibid.*, Art. 39, para. 1; emphasis added to 'normal modes'.
81. D. P. O'Connell, *supra* note 31, at 333.
82. *Ibid.*
83. J. N. Moore, 'The Regime of the Straits and the Third United Nations Conference on the Law of the Sea', *A.J.I.L.* 77–121, at 89 (1980).
84. Henry A. Kissinger, *International Law, World Order and Human Progress*, text distributed at meeting, 11 August 1975, at 7; see also text in 73 *Dep't State Bulletin* 353–62 (1975).
85. *Ibid.*,
86. For a discussion of the U.S. position on transit passage, see Jon H. Van Dyke (ed.), *Consensus and Confrontation: The United States and the Law of the Sea Convention*, 576 pages, at 292–311 (1985).
87. B. A. Harlow, 'Comment', 46 *Law and Contemporary Problems* 125–35, at 128 (1983).
88. Bernard Oxman, *supra* note 86, at 296.
89. David Colson, assistant legal adviser, in taped interview with C.B.C. reporter Bill Glaister, broadcast on 13 June 1985.
90. Statement in the House of Commons by Secretary of State for External Affairs Joe Clark, 10 Sept. 1985, reproduced in *Statement* Series 85/49, 4.
91. *Can. H. C. Debates*, at 39 (24 oct. 1969).
92. R.S.C. 1970, c. 2 (1st Supp.), s. 3.
93. *Ibid.*, s. 6.
94. *Ibid.*, s. 12.
95. *Ibidl.*

96. *Ibid.*, s. 23.
97. For a description of the system, see Transport Canada, *Offshore Vessel Traffic Management Systems Operations Standards (Interim), NORDREG CANADA*, 31 Aug. 1979, T.P. 1526, Element T-2.
98. 1982 Convention, Art. 234.
99. See 'La contribution du Canada au développement du droit international pour la protection du milieu marin: le cas spécial de l'Arctique', 11 *Etudes internationales* 441–66 (1980). It should be stated, however, that the United States still appears to consider the Canadian legislation as inconsistent with Article 234 of the Convention. See David A. Colson in J. M. Van Dyke (ed.), *Consensus and Confrontation: the United States and the Law of the Sea Convention*, at 424 (1985).
100. See *supra* note 1, at 119–20.
101. *Supra* note 90. In its Fourth Report, dated 14 February 1986, the Standing Committee of the House of Commons on External Affairs and National Defence expressed 'its support for the government's decision to build a class-8 icebreaker', at 32.
102. This was suggested in 1983 by the Senate Standing Committee on Foreign Affairs; see *Canada's Maritime Defence: Report of the Sub-Committee on National Defence*, at 92 (1983).
103. *Ibid.*, at 52.
104. *Supra* note 102, at 35.
105. *Supra* note 90.
106. See Exchange of Notes and Memorandum of Understanding Constituting an Agreement between the Government of Canada and the Government of the United States of America on the Modernization of the North American Air Defence System, Quebec City, March 18, 1985.
107. *Supra* note 102, at 33–4.
108. *Ibid.*, at 39.
109. *Supra* note 90, at 6.
110. This suggestion was previously made by this writer in 'The Legal Regime of the Arctic: Some Outstanding Issues', 34 *International Journal* 742–99, at 792–3 (1984).
111. *Supra* note 106, at 5.
112. *Supra* note 90, at 4.
113. Canada (Federal Environmental Assessment review office), *Final Report of the Environmental Assessment Panel*, at 96 (July 1984).
114. The Government is presently introducing related legislation, Bill C-75. This amendment to the Arctic Waters Pollution Prevention Act would permit the adoption of regulations establishing 'Vessel Traffic Services Zones', within existing shipping safety control zones, necessitating a traffic clearance for a ship to navigate in a Traffic Services Zone (Article 635.16 and 635.17).
115. Canada, Standing Committee on Foreign Affairs, Sub-Committee on National Defence, *Proceedings*, Issue No. 35, at 18 (23 Nov. 1982).
116. *Supra* note 102, at 50.
117. See Table 1 in Howard Hume, *A Comparative Study of the Northern Sea Route and the Northwest Passage*, Master's thesis in polar studies, Cambridge University, 15 June 1984, at 22.
118. See *Supra* note 90, at 2. In 1981, Japan was already inquiring from Canada as to its icebreaker technology and requirements for navigation in the Arctic waters.
119. See *supra* note 28, at 22–3.
120. *Supra* note 90, at 2.
121. D. S. Macdonald, *White Paper on Defence – Defence in the 70s*, at 18 (1971).
122. *Ibid.*

123. *Ibid.*
124. *Supra* note 102, at 51.
125. *Ibid.*
126. *Ibid.*
127. See Richard Halloran, 'Navy trains to battle Soviet Submarines in Arctic', *New York Times*, 19 May 1983, in Ron Purver, 'Security and Arms Control at the Poles', 39 *International Journal* 888–910, at 896 (1984).
128. On this point see generally W. Harriet Critchley, 'Polar Deployment of Soviet Submarines', 39 *International Journal* 828–87 (1984).
129. On this point, see in particular Howard Hume, *supra* Note 117, at 69, and John Gellner, 'Canada Already Has a Plan for the Arctic', *Globe and Mail*, 19 Sept. 1985, at 7.

General conclusion

Whenever appropriate, a summary has been made at the end of chapters, so that a general conclusion might not be absolutely necessary. However, for greater clarity and convenience, a few concluding remarks on the four parts of the book seem advisable. Even at the risk of redundancy, a conclusion is formulated on the main points developed in each of the four parts. The key words of the various chapter and section headings are retained so that the corresponding developments in the text may be easily located.

Part 1. The sector theory

The sector theory has no legal validity as a source of title or State jurisdiction in the Arctic. This applies to land and, *a fortiori*, to maritime areas. The conclusion follows from a study of three possible legal bases for the theory: the boundary treaties of 1825 and 1867, the doctrine of contiguity and custom.

Boundary treaties of 1825 and 1867 The meridians described in the 1825 Treaty between Great Britain and Russia, as well as the 1867 Treaty between the United States and Russia, were used only as a convenient geographical device to delimit territorial possessions. Although the meridians were described as extending as far as the Frozen Ocean, the Parties made it abundantly clear in both instances that the subject matter of the agreements was land only, and not land and sea.

Contiguity Subject to minor exceptions relating to certain islands in a territorial sea or forming an integral part of an Archipelago, the doctrine of contiguity does not constitute an adequate legal basis for the acquisition of sovereignty over land or maritime areas. Consequently, contiguity cannot serve as a valid basis for the sector theory.

Custom An examination of State practice in both the Arctic and the Antarctic reveals that the sector theory has not developed into a principle of customary law, either general or regional, and cannot serve as a root of title. Although Canada and the Soviet Union have relied on the theory in the Arctic, such reliance has been intermittent – certainly for Canada – and often ambiguous as to whether it envisaged land, sea, ice, continental shelf or air space. In addition, the other three Arctic States have never relied on the theory and the United States and Norway have always opposed it. In the Antarctic, all of the claims are described by the use of meridians but none expressly extend to the apex of the sectors of the South Pole; furthermore, Norway's claim is formulated in terms only of the coast and an undefined hinterland. Also of significance is the rejection by both the United States and the Soviet Union of all the Antarctic claims. Consequently, it is obvious that the sector theory is not the result of a settled practice, general or regional, which can be considered legally binding as a mode for the acquisition of territorial or maritime sovereignty.

Possible uses of sector theory and meridians In spite of its lack of validity as a legal root of title, the sector theory might prove to be a most convenient method for the delimitation of various forms of State juris-diction, including sovereignty. In the Arctic, for instance, States bordering the Arctic Ocean could agree to use sector lines to delimit their respective areas of responsibility in the implementation of their obligation under the Law of the Sea Convention to protect and preserve the marine environment.[1] They could also have recourse to the same mode of delimitation to coordinate their scientific research policies in that ocean.[2] In the same way, Arctic States could use sector lines to provide a control service for transpolar flights above a certain latitude. With the approval of the International Civil Aviation Organization, Canada began providing such a service for the sector of the Arctic Ocean between the 141st and 60th meridians in April 1970.[3] Finally, Canada might be permitted to follow the 141st meridian in the delimitation of its continental shelf in the Beaufort Sea because of its long-standing use for that purpose and a possible acquiescence in that use by the United States. This is all the more possible, given that the land boundary in the 1825 Treaty followed the 141st meridian and 'the continuation in the seaward direction of the land frontier' is one of the accepted methods of conti-nental shelf delimitation.[4] It must be well understood, however, that reliance by Canada on the 141st meridian would not be based on the sector theory.

Part 2. Historic waters
It is highly doubtful that Canada could succeed in proving that the waters of the Canadian Arctic Archipelago are historic internal waters over which it has complete sovereignty.

Status of historic waters Although the 1958 Territorial Sea Convention and the 1982 Law of the Sea Convention seem to envisage the possibility of an historic title resulting in territorial waters, customary law as applied by the International Court in the *Fisheries Case* of 1951 makes it clear that historic waters have the status of internal waters.

Role of historic waters The role of historic waters in international law has considerably diminished since the adoption of a 24-mile closing line for bays and the introduction of the straight baseline system for the measurement of the territorial sea.

Requirements of historic waters The requirements for the proof of historic waters are stringent and involve exclusive control and long usage by the claimant State as well as acquiescence by foreign States, particularly those clearly affected by the claim. The vital interests of the claimant State may be taken into account in the general appraisal of the claim. A claim of historic title to a maritime area is one in derogation of the standard rules for the acquisition of sovereignty over such area and imports a special burden of proof on the part of the claimant State.

Practice of Arctic States In 1951, Norway invoked historic fishing rights in some of the enclosed areas, along with the straight baseline system, in support of its claim of internal waters. In 1957, the Soviet Union claimed the Bay of Peter the Great, which is 108 miles wide, as an historic bay. In 1984, the Soviet Union claimed Penzhinskaya Inlet as historic, closing it with a 38.2 mile line and, in 1985, it added three bodies of water as historic internal waters: the White Sea, Cheshskaya Bay and Baydaratskaya Bay. The closing lines measure 84.4, 44 and 62.5 miles respectively. It also expressly provides for historic seas and straits in its legislation, but does not appear to have officially claimed any such sea or strait. The United States has claimed as historic a few small bays about 10 miles wide, but has made no claim of historic title to other maritime areas. It has strongly protested the Soviet Union's claim to the Bay of Peter the Great. Denmark has never made any claim of historic waters. Canada has claimed several bays as historic and, since 1973 at least, claims all the waters of its Arctic Archipelago as historic internal waters.

Canada's Case Canada's claim that it has an historic title to the waters of the Canadian Arctic Archipelago does not meet the stringent requirements of international law. On the positive side, it may be stated that virtually all of those waters were discovered by British explorers before the transfer of 1880 and most of them were explored and patrolled by Canada after that date. However, not only were the takings of possession by Great Britain and Canada limited to land and islands, but Canada has not always demonstrated the exclusive control required for the acquisition and maintenance of sovereignty over maritime areas. In addition, the acquiescence of other States, particularly that of the United States, is not sufficiently present. On the contrary, in 1970 the United States protested Canada's Arctic Waters Pollution Prevention Act which applied, not only to a 100-mile zone north of the mainland and the Archipelago, but to all the waters of the Archipelago itself. Canada's partial withdrawal of its acceptance of the International Court's jurisdiction, to prevent the testing of the international validity of this legislation relating to marine areas adjacent to the coast of Canada, could be interpreted as implying a doubt as to the precise legal status of the waters of the Archipelago. More important still is Canada's explanation of the legal effect of the accompanying piece of legislation extending its territorial waters from 3 to 12 miles, namely that it created a second gateway of territorial waters in the Northwest Passage, that is in Barrow Strait. These form part of the waters claimed to be internal since 1973. At the same time, the NORDREG reporting and clearance system, begun in 1977, is only 'voluntary', and foreign ships are so advised by the Coast Guard in its Notices to Mariners.[5] In these circumstances, Canada could hardly make a case for historic internal waters.

Part 3. Straight baselines
Canada has validly drawn straight baselines around the Canadian Arctic Archipelago, with the result that the enclosed waters, including those of the Northwest Passage, have the status of internal waters. Under customary law, as applied in the *Fisheries Case*, the enclosed waters are not subject to the right of innocent passage. They would have been under the 1958 Territorial Sea Convention and the 1982 Law of the Sea Convention.

Geographical requirements The geography required for the applicability of straight baselines, as expressed in the *Fisheries Case*, is that the coast be indented or 'bordered by an Archipelago such as the (Norwegian) skjaergaard'. This coastal Archipelago is comprised of some 120,000 insular formations carved out of a mainland coast, broken by large and deeply indented fjords, obliterating any clear dividing line between the mainland and the sea. Some of the islands are located some 60 miles from the nearest peninsula on the mainland. This customary law requirement for coastal Archipelagos, exemplified by the Norwegian skjaergaard, was incorporated in both the Territorial Sea Convention of 1958 and the Law of the Sea Convention of 1982. Although expressed in those Conventions as requiring a 'fringe of islands along the coast in its immediate vicinity', no difference seems to have been intended.

Legal nature In the *Fisheries Case*, the International Court disagreed with the view of the United Kingdom that straight baselines constituted an exceptional system and held that they simply represented 'the application of general international law to a specific case'.[6] The Court considered the straight baseline system as a particular application of the rule of the low-water mark from which to measure the breadth of the territorial sea. In spite of this clear holding, the provisions for straight baselines in the 1958 and 1982 Conventions are expressed in the form of exceptions to the low-water line along the coast. It is not certain that the formal difference was intended to be a substantive difference as well.

Mode of application In the actual drawing of straight baselines, three criteria or guidelines are involved: two geographic criteria which are compulsory and one economic criterion which is optional. The first compulsory criterion is that straight baselines must not depart to any appreciable extent from the *general direction* of the coast. The second compulsory criterion is that there must be a sufficiently *close link* between the land domain and the newly enclosed waters to subject the latter to the regime of internal waters. The third and optional criterion is that, in the establishment of particular lines, account may be taken of certain *economic interests* peculiar to the regions concerned providing that their reality and importance are clearly evidenced by long usage.

Length There is no maximum length for straight baselines enclosing coastal Archipelagos and drawn in conformity with the required criteria.

Consolidation of title to enclosed waters Long usage may be invoked to consolidate a title to sea areas as a secondary basis, in addition to a primary basis such as the straight baseline system. History is then used not as the essential or main ground for the title – this would be the function of an historic title – but as a subsidiary ground confirming and consolidating the title resulting from the enclosure of certain sea areas by straight baselines. In the *Fisheries Case*, the historic and hunting rights enjoyed by the local inhabitants were held to confirm the validity of the straight baseline across the waters of the Lopphavet by way of consolidation of title to those waters.

Since a consolidation – unlike an historical title – does not constitute a derogation from the standard rules for the acquisition of maritime sovereignty, there is no special burden of proof and the requirements are less stringent. These requirements are: some exercise of State authority, long usage and a general toleration by foreign States. The claimant State may also take into account certain vital interests, such as those resulting from its duty to protect the vital economic needs of its local inhabitants.

State practice The great majority of States have either actually used straight baselines to measure the breadth of their territorial sea or have adopted enabling legislation. Four of the five Arctic States (Norway, Denmark, the Soviet Union and Canada) have made use of straight baselines. The fifth Arctic State (United States) has not used straight baselines and has no enabling legislation. However, it has recognized the applicability of the straight baseline system to highly irregular and fragmented coasts.

Geography of the Canadian Arctic Archipelago The Archipelago is a huge labyrinth of islands and headlands of various shapes and sizes, fused together by a solid ice cover most of the year which renders it often impossible to distinguish between land and sea. Those land formations constitute a single unit bordering the northern coast of Canada and forming an integral part of it. The physical characteristics of this ensemble prohibit any attempt to follow the sinuosities of the mainland coast or of the islands in the measurement of the territorial sea, and command the utilization of straight baselines.

Canada's baselines The newly established straight baselines meet the two compulsory geographic criteria, in that the baselines follow the *general direction* of the outer line of the Archipelago, which really constitutes the Canadian coast; there also exists a *close link* between the land and the sea, the ratio being 1 to 0.822. By comparison, the ratio for the Norwegian Archipelago is 1 to 3.5. In addition, the validity of the straight baseline across Amundsen Gulf and Lancaster Sound is reinforced by the *economic interests* of the local Inuit population whose fishing, hunting and trapping can be traced back to time immemorial.

Length Although there is no legal limit to the length of straight baselines for this type of Archipelago, an effort has been made to restrict such length, with the result that the longest baseline is 99.5 miles and the average length is less than 17 miles.

Consolidation of title to enclosed waters As an additional basis to the straight baseline system, Canada can rely on an historical consolidation of title to the waters of Lancaster Sound, Barrow Strait and Amundsen Gulf. This consolidation results from: (1) the exercise of legislative and administrative jurisdiction by Canada for about 80 years; (2) the immemorial usage of those waters by the Inuit; and (3) the general toleration of foreign States. Canada may also legitimately invoke its vital interests in Arctic waters, particularly its obligation to protect the unique marine environment, the vital needs of the Inuit and the security of its maritime territory.

Part 4. The Northwest Passage

The Northwest Passage has never been used for international navigation and is not an international strait. Since its enclosure by straight baselines, not even the right of innocent passage applies. However, should there be a sufficient use of the Passage by foreign commercial ships, the new right of 'transit passage' would apply and it could not be suspended.

Main routes There are five main routes in the Northwest Passage, two of which are potentially suitable for deep draft (20 metres) navigation. Of these two, only one is sufficiently free from polar ice to be used for shipping hydrocarbons from the Beaufort Sea area. Proceeding from east to west, this route lies across five bodies of water: Lancaster Sound, Barrow Strait, Viscount Melville Sound, Prince of Wales Strait and Amundsen Gulf.

All the main bodies of water of the Northwest Passage were discovered by the British expeditions before the transfer of the Arctic Islands to Canada in 1880 and were then explored and patrolled by Canada. Thus far, only 45 complete transits of the Northwest Passage have taken place, 29 of them by Canadian ships. Of the 16 foreign crossings, 11 were American, one Norwegian, one Dutch, one Japanese, one Bahamian and one Liberian. All of the foreign transits were accomplished with Canada's consent or acquiescence. Of the 45 transits, only three could perhaps be characterized as having been effected by commercial ships: the tanker

Manhattan in 1969 and the cruise ships *Lindblad Explorer* in 1984 and *World Discoverer* in 1985.

Future use Vast hydrocarbon resources have been located in the Canadian Arctic Archipelago and in the Beaufort Sea area. Both the United States and Canada are determined to attain energy self-sufficiency as soon as possible and have intensified their drilling activities. Technology for production and year-round navigation has been developed, and it appears certain that the Northwest Passage will eventually be used for the shipping of hydrocarbons. Considerable uncertainty remains as to when such shipping will begin, but a probable date would appear to be in the late 1990s. The implementation of a pilot project has already started with the shipping of 100,000 barrels of light crude oil from Cameron Island (close to the western end of the Passage) in August 1985 and the project calls for an eventual increase to 50,000 barrels a day with large ice-breaking tankers.

Present legal status As was made clear by the International Court in the *Corfu Channel Case* of 1949, a substantial degree of use by foreign shipping is necessary before a strait may be considered in international law as one 'used for international navigation'. Although a somewhat lesser degree of use might be required in a remote region such as the Arctic, virtually none has taken place across the Northwest Passage and it cannot possibly be considered an international strait. In addition, since the newly established straight baselines came into effect on January 1, 1986, the Northwest Passage has become a national sea route. Its waters are strictly internal waters, without any right of innocent passage. However, it probably is in Canada's best interest to permit the passage of foreign ships on certain conditions.

Future legal status In spite of its present legal status as a national sea route, it is still possible for the Northwest Passage to become an international strait. If Canada were not to take the necessary measures to exercise effective control over foreign ships using the Passage, internationalization might well result, with the consequent right of transit passage in favour of all ships. To prevent this, Canada has already announced certain measures in addition to the existing one relating to the protection of the marine environment. In particular, Canada has decided to build a Class 8 icebreaker, increase its surveillance overflights and insure a naval presence in the Eastern Arctic waters. The Class 8 icebreaker, which could be so designed as to permit the installation of weapons systems, will enable Canada to insure a certain presence in Route 1 (see Figure 18), through Prince of Wales Strait on a year-round basis. Only one such icebreaker will not suffice for eventual year-round surveillance and control of foreign shipping and consideration should be given to the building of a Class 10 icebreaker. This is required by Canada's own regulations for year-round navigation in M'Clure Strait. In addition, and because of the activities of foreign submarines under the ice pack of Arctic waters, possibly extending to those of its Arctic Archipelago, Canada should develop the most effective submarine detection capability for the ice-covered areas of its Arctic waters in general and those of the Northwest Passage in particular.

Notes to the General conclusion

1. See Article 123 of the 1982 Convention on the cooperation of States bordering enclosed or semi-enclosed seas, in particular paragraph (b).
2. *Ibid.*, para. c.
3. See K. R. Greenaway and M. Dunbar, 'Aviation in the Arctic Islands', in M. Zaslow (ed), *A Century of Canada's Arctic Islands* 78–92, at 88 (1981).
4. This is one of four methods of delimitation considered by the Committee of Experts of the International Law Commission and mentioned by the International Court; see *North Sea Continental Shelf Cases* (1969) I.C.J. rep., at 34.
5. See Canadian Coast Guard, *Notices to Mariners* 1 to 38, Annual Edition, at 179 (1984).
6. [1951] I.C.J. Rep., at 131.

Appendix A

The 1825 Boundary Treaty, Great Britain and Russia

Convention between Great Britain and Russia concerning limits of their respective possessions on the north-west coast of America and the navigation of th Pacific Ocean, signed at St Petersburg, 16(28) February 1825.

75 Consolidated Treaty Series, 96.

In the Name of the Most Holy and Undivided Trinity.

(Translation.)

His Majesty The King of the United Kingdom of Great Britain and Ireland, and His Majesty The Emperor of all the Russias, being desirous of drawing still closer the Ties of good Understanding and Friendship which unite them, by means of an Agreement which may settle, upon the basis of reciprocal convenience, different points connected with the Commerce, Navigation, and Fisheries of their subjects on the Pacific Ocean, as well as the limits of their respective Possessions on the North-West Coast of America, have named Plenipotentiaries to conclude a Convention for this purpose, that is to say:—His Majesty The King of the United Kingdom of Great Britain and Ireland, The Right Honourable Stratford Canning, a Member of His said Majesty's Most Honourable Privy Council, &c. And His Majesty The Emperor of all the Russias, The Sieur Charles Robert Count de Nesselrode, His Imperial Majesty's Privy Councillor, a Member of the Council of the Empire, Secretary of State for the Department of Foreign Affairs, &c. ; and the Sieur Pierre de Poletica, His Imperial Ma-

Au Nom de la Très Sainte et Indivisible Trinité.

Sa Majesté le Roi du Royaume Uni de La Grande Bretagne et de l'Irlande, et Sa Majesté l'Empereur de toutes les Russies, désirant resserrer les liens de bonne intelligence et d'amitié qui les unissent, au moyen d'un accord qui régleroit, d'après le principe des convenances réciproques, divers points relatifs au Commerce, à la Navigation, et aux Pêcheries de leurs Sujets sur l'Océan Pacifique, ainsi que les limites de leurs Possessions respectives sur la Côte Nord-Ouest de l'Amérique, ont nommé des Plénipotentiaires pour conclure une Convention à cet effet, savoir : —Sa Majesté le Roi du Royaume Uni de La Grande Bretagne et de l'Irlande, Le Très Honorable Stratford Canning, Conseiller de Sa dite Majesté en Son Conseil Privé, &c. Et Sa Majesté l'Empereur de toutes les Russies, le Sieur Charles Robert Comte de Nesselrode, Son Conseiller Privé actuel, Membre du Conseil de l'Empire, Secrétaire d'Etat dirigeant le Ministère des Affaires Etrangères, &c.; et le Sieur Pierre de Poletica, Son Conseiller d'Etat actuel, &c. Lesquels Plénipotentiaires, après s'être communiqué leurs Pleinpouvoirs respectifs, trouvés en

jesty's Councillor of State, &c. Who, after having communicated to each other their respective Full Powers, found in good and due form, have agreed upon and signed the following Articles:—

I. It is agreed that the respective Subjects of The High Contracting Parties shall not be troubled or molested, in any part of the Ocean, commonly called the Pacific Ocean, either in navigating the same, in fishing therein, or in landing at such Parts of the Coast as shall not have been already occupied, in order to trade with the Natives, under the restrictions and conditions specified in the following Articles.

II. In order to prevent the Right of navigating and fishing exercised upon the Ocean by the Subjects of The High Contracting Parties, from becoming the pretext for an illicit Commerce, it is agreed that the Subjects of His Britannick Majesty shall not land at any place where they may be a Russian Establishment, without the permission of the Governor or Commandant; and, on the other hand, that Russian Subjects shall not land, without permission, at any British Establishment on the North-West Coast.

III. The line of demarcation between the Possessions of the High Contracting Parties, upon the Coast of the Continent, and the Islands of America to the North-West, shall be drawn in the manner following:—

bonne et due forme, ont arrêté et signé les Articles suivans:—

I. Il est convenu que dans aucune partie du Grand Océan, appelé communément Océan Pacifique, les Sujets respectifs des Hautes Puissances Contractantes ne seront ni troublés, ni gênés, soit dans la navigation, soit dans l'exploitation de la pêche, soit dans la faculté d'aborder aux Côtes, sur des Points qui ne seroient pas déjà occupés, afin d'y faire le commerce avec les Indigènes, sauf toutefois les restrictions et conditions déterminées par les Articles qui suivent.

II. Dans la vûe d'empêcher que les droits de navigation et de pêche exercés sur le Grand Océan par les Sujets des Hautes Parties Contractantes, ne deviennent le prétexte d'un commerce illicite, il est convenu que les Sujets de Sa Majesté Britannique n'aborderont à aucun Point où il se trouve un Etablissement Russe, sans la permission du Gouverneur ou Commandant, et que, réciproquement, les Sujets Russes ne pourront aborder, sans permission, à aucun Etablissement Britannique sur la Côte Nord-Ouest.

III. La ligne de démarcation entre les Possessions des Hautes Parties Contractantes sur la Côte du Continent et les Iles de l'Amérique Nord-Ouest, sera tracée ainsi qu'il suit:—

Commencing from the Southernmost Point of the Island called *Prince of Wales* Island, which Point lies in the parallel of 54 degrees 40 minutes North latitude, and between the 131st and the 133d degree of West longitude (Meridian of Greenwich), the said line shall ascend to the North along the Channel called *Portland Channel,* as far as the Point of the Continent where it strikes the 56th degree of North latitude; from this last-mentioned Point, the line of demarcation shall follow the summit of the mountains situated parallel to the Coast, as far as the point of intersection of the 141st degree of West longitude (of the same Meridian); and, finally, from the said point of intersection, the said Meridian Line of the 141st degree, in its prolongation as far as the Frozen Ocean, shall form the limit between the Russian and British Possessions on the Continent of America to the North-West.

IV. With reference to the line of demarcation laid down in the preceding Article, it is understood;

1st. That the Island called *Prince of Wales* Island shall belong wholly to Russia.

2nd. That wherever the summit of the mountains which extend in a direction parallel to the Coast, from the 56th degree of North latitude to the point of intersection of the 141st degree of West longitude, shall prove to be at the distance of more than ten marine leagues from the Ocean, the limit between the British Possessions

A partir du Point le plus méridional de l'Ile dite *Prince of Wales,* lequel Point se trouve sous la parallèle du 54me degré 40 minutes de latitude Nord, et entre le 131me et le 133me degré de longitude Ouest (Méridien de Greenwich), la dite ligne remontera au Nord le long de la passe dite *Portland Channel,* jusqu'au Point de la terre ferme où elle atteint le 56me degré de latitude Nord ; de ce dernier point la ligne de démarcation suivra la crête des montagnes situées parallèlement à la Côte, jusqu'au point d'intersection du 141me degré de longitude Ouest (même Méridien) ; et, finalement, du dit point d'intersection, la même ligne méridienne du 141me degré formera, dans son prolongement jusqu'à la Mer Glaciale, la limite entre les Possessions Russes et Britanniques sur le Continent de l'Amérique Nord-Ouest.

IV. Il est entendu, par rapport à la ligne de démarcation déterminée dans l'Article précédent ;

1. Que l'Isle dite *Prince of Wales* appartiendra toute entière à la Russie.

2. Que partout où la crête des montagnes qui s'étendent dans une direction parallèle à la Côte depuis le 56me degré de latitude Nord au point d'intersection du 141me degré de longitude Ouest, se trouveroit à la distance de plus de dix lieues marines de l'Océan, la limite entre les Possessions Britanniques et la lisière de Côte

and the line of Coast which is to belong to Russia, as above mentioned, shall be formed by a line parallel to the windings of the Coast, and which shall never exceed the distance of ten marine leagues therefrom.

V. It is moreover agreed, that no Establishment shall be formed by either of the Two Parties within the limits assigned by the two preceding Articles to the Possessions of the Other: consequently, British Subjects shall not form any Establishment either upon the Coast, or upon the border of the Continent comprised within the limits of the Russian Possessions, as designated in the two preceding Articles; and, in like manner, no Establishment shall be formed by Russian Subjects beyond the said limits.

VI. It is understood that the Subjects of His Britannick Majesty, from whatever Quarter they may arrive, whether from the Ocean, or from the interior of the Continent, shall for ever enjoy the right of navigating freely, and without any hindrance whatever, all the rivers and streams which, in their course towards the Pacific Ocean, may cross the line of demarcation upon the line of Coast described in Article 3 of the present Convention.

VII. It is also understood, that, for the space of ten Years from the signature of the present Convention, the Vessels of the Two Powers, or those belonging to Their respective Subjects, shall mutually be at liberty to frequent,

mentionnée ci-dessus comme devant appartenir à La Russie, sera formée par une ligne parallèle aux sinuosités de la Côte, et qui ne pourra jamais en être éloignée que de dix lieues marines.

V. Il est convenu en outre, que nul Etablissement ne sera formé par l'une des deux Parties dans les limites que les deux Articles précédens assignent aux Possessions de l'Autre. En conséquence, les Sujets Britanniques ne formeront aucun Etablissement soit sur la Côte, soit sur la lisière de terre ferme comprise dans les limites des Possessions Russes, telles qu'elles sont désignées dans les deux Articles précédens; et, de même, nul Etablissement ne sera formé par des Sujets Russes au delà des dites limites.

VI. Il est entendu que les Sujets de Sa Majesté Britannique, de quelque Côté qu'ils arrivent, soit de l'Océan, soit de l'intérieur du Continent, jouiront à perpétuité du droit de naviguer librement, et sans entrave quelconque, sur tous les fleuves et rivières, qui, dans leurs cours vers la Mer Pacifique, traverseront la ligne de démarcation sur la lisière de la Côte indiquée dans l'Article 3 de la présente Convention.

VII. Il est aussi entendu que, pendant l'espace de dix Ans, à dater de la signature de cette Convention, les Vaisseaux des deux Puissances, ou ceux appartenans à leurs Sujets respectifs, pourront réciproquement fréquenter, sans

without any hindrance whatever, all the inland Seas, the Gulfs, Havens, and Creeks on the Coast mentioned in Article 3, for the purposes of fishing and of trading with the Natives.

VIII. The Port of Sitka, or Novo Archangelsk, shall be open to the Commerce and Vessels of British Subjects for the space of ten Years from the date of the exchange of the Ratifications of the present Convention. In the event of an extension of this term of ten Years being granted to any other Power, the like extension shall be granted also to Great Britain.

IX. The above-mentioned liberty of Commerce shall not apply to the trade in spirituous liquors, in fire-arms or other arms, gunpowder or other warlike stores ; the High Contracting Parties reciprocally engaging not to permit the above-mentioned articles to be sold or delivered, in any manner whatever, to the Natives of the Country.

X. Every British or Russian Vessel navigating the Pacific Ocean, which may be compelled by storms or by accident, to take shelter in the Ports of the respective Parties, shall be at liberty to refit therein, to provide itself with all necessary stores, and to put to sea again, without paying any other than Port and Lighthouse dues, which shall be the same as those paid by National Vessels. In case, however, the Master of such Vessel should be under the necessity of disposing

entrave quelconque, toutes les Mers intérieures, les Golfes, Havres, et Criques sur la Côte mentionnée dans l'Article 3, afin d'y faire la pêche et le commerce avec les Indigènes.

VIII. Le Port de Sitka, ou Novo Archangelsk, sera ouvert au Commerce et aux Vaisseaux des Sujets Britanniques durant l'espace de dix Ans, à dater de l'échange des Ratifications de cette Convention. Au cas qu'une prolongation de ce terme de dix Ans soit accordée à quelque autre Puissance, la même prolongation sera également accordée à La Grande Bretagne.

IX. La susdite liberté de commerce ne s'appliquera point au trafic des liqueurs spiritueuses, des armes à feu, des armes blanches, de la poudre à canon, ou d'autres munitions de guerre ; les Hautes Parties Contractantes s'engageant réciproquement à ne laisser ni vendre, ni livrer, de quelque manière que ce puisse être, aux Indigènes du Pays, les articles ci-dessus mentionnés.

X. Tout Vaisseau Britannique ou Russe naviguant sur l'Océan Pacifique, qui sera forcé par des tempêtes, ou par quelque accident, de se réfugier dans les Ports des Parties respectives, aura la liberté de s'y radouber, de s'y pourvoir de tous les objets qui lui seront nécessaires, et de se remettre en mer, sans payer d'autres Droits que ceux de Port et de Fanaux, lesquels seront pour lui les mêmes que pour les Bâtimens Nationaux. Si, cependant, le Patron d'un tel navire se trouvoit dans la néces-

of a part of his merchandize in order to defray his expences, he shall conform himself to the Regulations and Tariffs of the Place where he may have landed.

XI. In every case of complaint on account of an infraction of the Articles of the present Convention, the Civil and Military Authorities of The High Contracting Parties, without previously acting or taking any forcible measure, shall make an exact and circumstantial Report of the matter to their respective Courts, who engage to settle the same, in a friendly manner, and according to the principles of justice.

XII. The present Convention shall be ratified, and the Ratifications shall be exchanged at London, within the space of six weeks, or sooner if possible.

In Witness whereof the respective Plenipotentiaries have signed the same, and have affixed thereto the Seal of their Arms.

Done at St. Petersburgh, the $\frac{\text{Twen v Eighth}}{\text{Sixteenth}}$ day of February, in the Year of our Lord one thousand eight hundred and twenty-five.

(L.S.)
STRATFORD CANNING.
(L.S.)
The COUNT de NESSELRODE.
(L.S.)
PIERRE DE POLETICA.

---

sité de se défaire d'une partie de ses marchandises pour subvenir à ses dépenses, il sera tenu de se conformer aux Ordonnances et aux Tarifs de l'Endroit où il aura abordé.

XI. Dans tous les cas de plaintes relatives à l'infraction des Articles de la présente Convention, les Autorités Civiles et Militaires des deux Hautes Parties Contractantes, sans se permettre au préalable ni voie de fait, ni mesure de force, seront tenues de faire un rapport exact de l'affaire et de ses circonstances à leurs Cours respectives, lesquelles s'engagent à la régler à l'amiable, et d'après les principes d'une parfaite justice.

XII. La présente Convention sera ratifiée, et les Ratifications en seront échangées à Londres dans l'espace de six semaines, ou plutôt si faire se peut.

En Foi de quoi les Plénipotentiaires respectifs l'ont signée, et y ont apposé le Cachet de leurs Armes.

Fait à St. Pétersbourg, le $\frac{\text{Vingt-huit}}{\text{Seize}}$ Février, de l'An de Grâce mil huit cent vingt-cinq.

(L.S.)
STRATFORD CANNING.
(L S.)
Le COMTE de NESSELRODE.
(L.S.)
PIERRE DE POLETICA.

75 Consolidated Treaty Series, 96.

Appendix B

The 1867 Boundary Treaty, United States and Russia

Convention ceding Alaska between Russia and the United States, signed at Washington, 30 March 1867.

Concluded March 30, 1867; ratification advised by the Senate April 9, 1867; ratified by the President May 28, 1867; ratifications exchanged June 20, 1867; proclaimed June 20, 1867.

ARTICLES.

I. Territory ceded; boundaries.	IV. Formal delivery.
II. Public property ceded.	V. Withdrawal of troops.
III. Citizenship of inhabitants; un- civilized tribes.	VI. Payment; effect of cession.
	VII. Ratification.

The United States of America and His Majesty the Emperor of all the Russias, being desirous of strengthening, if possible, the good understanding which exists between them, have, for that purpose, appointed as their Plenipotentiaries: the President of the United States, William H. Seward, Secretary of State; and His Majesty the Emperor of all the Russias, the Privy Counsellor Edward de Stoeckl, his Envoy Extraordinary and Minister Plenipotentiary to the United States.

And the said Plenipotentiaries, having exchanged their full powers, which were found to be in due form, have agreed upon and signed the following articles:

ARTICLE I.

His Majesty the Emperor of all the Russias agrees to cede to the United States, by this convention, immediately upon the exchange of the ratifications therof, all the territory and dominion now possessed by his said Majesty on the continent of America and in the adjacent islands, the same being contained within the geographical limits herein set forth, to wit: The eastern limit is the line of demarcation between the Russian and the British possessions in North America, as established by the convention between Russia and Great Britain, of February 28–16, 1825, and described in Articles III and IV of said convention, in the following terms:

"Commencing from the southernmost point of the island called Prince of Wales Island, which point lies in the parallel of 54 degrees 40 minutes north latitude, and between the 131st and 133^d degree of west longitude, (meridian of Greenwich,) the said line shall ascend to the north along the channel called Portland channel, as far as the point of the continent where it strikes the 56th degree of north latitude; from this last mentioned point, the line of demarcation shall follow the summit of the mountains situated parallel to the coast as far as the point of intersection of the 141st degree of west longitude, (of the same meridian;) and finally, from the said point of intersection, the said meridian line of the 141st degree, in its prolongation as far as the Frozen ocean.

"IV. With reference to the line of demarcation laid down in the preceding article, it is understood—

"1st That the island called Prince of Wales Island shall belong wholly to Russia," (now, by this cession, to the United States.)

"2^d. That whenever the summit of the mountains which extend in a direction parallel to the coast from the 56th degree of north latitude to the point of intersection of the 141st degree of west longitude shall prove to be at the distance of more than ten marine leagues from the ocean, the limit between the British possessions and the line of coast which is to belong to Russia as above mentioned (that is to say, the limit to the possessions ceded by this convention) shall be formed by a line parallel to the winding of the coast, and which shall never exceed the distance of ten marine leagues therefrom."

The western limit within which the territories and dominion conveyed, are contained, passes through a point in Behring's straits on the parallel of sixty-five degrees thirty minutes north latitude, at its intersection by the meridian which passes midway between the islands

of Krusenstern, or Ignalook, and the island of Ratmanoff, or Noonarbook, and proceeds due north, without limitation, into the same
Frozen Ocean. The same western limit, beginning at the same initial
point, proceeds thence in a course nearly southwest, through Behring's
straits and Behring's sea, so as to pass midway between the northwest
point of the island of St. Lawrence and the southeast point of Cape
Choukotski, to the meridian of one hundred and seventy-two west
longitude; thence, from the intersection of that meridian, in a southwesterly direction, so as to pass midway between the island of Attou
and the Copper island of the Kormandorski couplet or group, in the
North Pacific ocean, to the meridian of one hundred and ninety-three
degrees west longitude, so as to include in the territory conveyed the
whole of the Aleutian islands east of that meridian.

ARTICLE II.

In the cession of territory and dominion made by the preceding
article, are included the right of property in all public lots and
squares, vacant lands, and all public buildings, fortifications, bar-
racks, and other edifices which are not private individual property.
It is, however, understood and agreed, that the churches which have
been built in the ceded territory by the Russian government, shall
remain the property of such members of the Greek Oriental Church
resident in the territory, as may choose to worship therein. Any
Government archives, papers, and documents relative to the territory
and dominion aforesaid, which may now be existing there, will be left
in the possession of the agent of the United States; but an authenti-
cated copy of such of them as may be required, will be, at all times,
given by the United States to the Russian government, or to such
Russian officers or subjects as they may apply for.

ARTICLE III.

The inhabitants of the ceded territory, according to their choice,
reserving their natural allegiance, may return to Russia within three
years; but if they should prefer to remain in the ceded territory, they,
with the exception of uncivilized native tribes, shall be admitted to
the enjoyment of all the rights, advantages and immunities of citizens
of the United States, and shall be maintained and protected in the
free enjoyment of their liberty, property and religion. The uncivilized tribes will be subject to such laws and regulations as the United
States, may from time to time, adopt in regard to aboriginal tribes
of that country.

ARTICLE IV.

His Majesty, the Emperor of all the Russias shall appoint, with
convenient despatch, an agent or agents for the purpose of formally
delivering to a similar agent or agents appointed on behalf of the
United States, the territory, dominion, property, dependencies and
appurtenances which are ceded as above, and for doing any other act
which may be necessary in regard thereto. But the cession, with the
right of immediate possession, is nevertheless to be deemed complete
and absolute on the exchange of ratifications, without waiting for
such formal delivery.

ARTICLE V.

Immediately after the exchange of the ratifications of this convention, any fortifications or military posts which may be in the ceded
territory, shall be delivered to the agent of the United States, and
any Russian troops which may be in the Territory shall be withdrawn
as soon as may be reasonably and conveniently practicable.

ARTICLE VI.

In consideration of the cession aforesaid, the United States agree to pay at the Treasury in Washington, within ten months after the exchange of the ratifications of this convention, to the diplomatic representative or other agent of his Majesty the Emperor of all the Russias, duly authorized to receive the same, seven million two hundred thousand dollars in gold. The cession of territory and dominion herein made is hereby declared to be free and unincumbered by any reservations, privileges, franchises, grants, or possessions, by any associated companies, whether corporate or incorporate, Russian or any other, or by any parties, except merely private individual property-holders; and the cession hereby made, conveys all the rights, franchises, and privileges now belonging to Russia in the said territory or dominion, and appurtenances thereto.

ARTICLE VII.

When this convention shall have been duly ratified by the President of the United States, by and with the advice and consent of the Senate, on the one part, and on the other by His Majesty the Emperor of all the Russias, the ratifications shall be exchanged at Washington within three months from the date hereof, or sooner, if possible.

In faith whereof, the respective plenipotentiaries have signed this convention, and thereto affixed the seals of their arms.

Done at Washington, the thirtieth day of March in the year of our Lord one thousand eight hundred and sixty-seven.

[SEAL.]   EDOUARD DE STOECKL,
[SEAL.]   WILLIAM H. SEWARD,

134 Consolidated Treaty Series, 332.

Selected bibliography

This bibliography is divided into four parts, representing the main questions examined in the book. Wherever appropriate, emphasis is placed on the more recent sources in the selection. The bibliography does not include a number of brief statements referred to in the text, such as those in the House of Commons and Senate on the sector theory.

Some bibliographical sources

Arctic Institute of North America, Arctic Bibliography (1953–72), 16 volumes. Montreal and London: McGill–Queen's University Press.

Breitfuss, L., Chronological list of 'the most important voyages and events in the Arctic since the beginning of our history', *The Arctic*, at 126–83 (1939).

Cooke A. and Holland, C., *The Exploration of Northern Canada, 500–1920, A. Chronology*, 549 pages (1978). Toronto: Arctic History Press.

Corley, N. T., 'Bibliography of Books Printed before 1800', 19 *Arctic* 77–98 (1966).

Corley, N. T., *Polar and Cold Regions Library Resources*, a directory covering twenty countries (1975). Ottawa: Defence Scientific Information Service.

Dartmouth College Library, *Dictionary Catalogue of the Stefansson Collection on Polar Regions* (1967). Boston: G. K. Hall & Co.

Dunbar, M. J., *Marine Transportation and High Arctic Development: A Bibliography*, 162 pages (1979). Ottawa: Canadian Arctic Resources Committee.

Dutilly, A. A., *Bibliography of Bibliographies on the Arctic* (1945). Washington: Catholic University of America Press.

Holland C., *Manuscripts in the Scott Polar Research Institute, Cambridge, England, A Catalogue*. New York and London: Garland Publishing Inc., 815 pages (1982).

Sanderson, D. and Smith, G., *Ships Navigating in Ice, A Selected Bibliography*, 129 pages (1982). Ottawa: Transport Canada.

Schmidt, R. W., *Arctic and Subarctic Transportation: A Tentative Bibliography* (1949). Maxwell Air Force Base, Air University.

Smith, Gordon W., 'Bibliography' in *The Historical and Legal Background of Canada's Arctic Claim*, at 465–96 (1952). New York: Doctoral dissertation, Faculty of Political Science, Columbia University.

Stairs, A., *Canadian Administrative and Other Activity in the North c. 1920–60*, selected list of Articles from leading magazines and journals, 29 typewritten pages (1972). Ottawa: Department of Indian Affairs and Northern Development.

Taylor, A., *Arctic Blue Books*, British Parliamentary Papers on Exploration of the Canadian North, 10 volumes, plus Index reproduced from Vol. 7 of *Arctic Bibliography*.

Wiktor, C. L. *Canadian Bibliography of International law*, 767 pages, at 197–207 (1984). Toronto: University of Toronto Press.

1 Sector theory

Auburn, F. M., *The Ross Dependency*, 91 pages, at 24–30 (1972).

Beal, Allan M., 'Comment on Commander Partridge's . . . Acceptance of the Sector Principle', 88 *U.S. Naval Institute proceedings* 115–18 (1962).

Bériault, Yvon, *Les problèmes politiques du nord canadien*, 201 pages, at 43–52 (1942).

Breitfuss, L., 'Territorial Division of the Arctic', 8 *Dalhousie Review* 456–70 (1929).

Canada, *The Yukon Territory Act*, 1898, Statutes of Canada, Chapter 6.

Canada and Norway, 'Exchange of Notes', 8 Aug. 1930 and 5 Nov. 1930, *Canada Treaty Series*, No. 17.

Cooper, J. C., 'Airspace Rights over the Arctic', 3 *Air Affairs* 516–40 (1950).

Dollot, R., 'Le Droit international des espaces polaires', 75 *Recueil des Cours*, Tôme II, 121–72, at 130–5 (1949).

Franklin, C. M. and McClintock, V. C., 'The Territorial Claims of Nations in the Arctic: An Appraisal', 5 *Oklahoma Law Review* 37–48 (1952).

Gidel, G., *Aspects juridiques de la lutte pour l'Antarctique*, Académie de Marine, Paris, at 29–46 (1948).

Gidel, G., *Les Poles et le Droit des Gens*, Cours à l'Institut des Hautes Etudes Internationales, Université de Paris (1950).

Hyde, C. C., 'Acquisition of Sovereignty over Polar Areas', 19 *Iowa Law Review* 286–94 (1934).

Lakhtine, W. L., 'Rights over the Arctic', 24 *American Journal of International Law* 703–17 (1930).

Lauterpacht, H., 'Sovereignty over Submarine Areas', 27 *British Yearbook of International Law* 376–433 (1950).

Miller, D. H., 'Political Rights in the Arctic', 4 *Foreign Affairs* 47–60 (1925).

Miller, D. H., 'Political Rights in the Polar Regions' in W. L. G. Joerg (ed.), *Problems of Polar Research* 235–50, 479 pages (1928).

Nicholson, N. L., *The Boundaries of Canada, its Provinces and Territories*, 142 pages, at 36–44 (1964).

Partridge, B., 'The White Shelf: A Study of Arctic Ice Jurisdiction', 87 *U.S. Naval Institute Proceedings* 51–7 (1961).

Pearson, L. B., 'Canada Looks "Down North" ', 24 *Foreign Affairs* 638–47 (1946).

Pearson, L. B., 'Canada's Northern Horizon', 31 *Foreign Affairs* 581–91 (1953).

Pharand, D., *La théorie des secteurs dans l'Arctique à l'égard du droit international*, doctoral dissertation at Faculty of Law, University of Paris, 166 pages (1955).

Plischke, E., 'Transpolar Aviation and Jurisdiction over Arctic Airspace', 6 *American Political Science Review* 999–1013 (1943).

Plischke, E., *Jurisdiction in the Polar Regions*, Ph.D. dissertation, Department of History and International Relations, Clark University, Massachusetts (1943).

Reeves, J. S., 'Editorial Comment: Antarctic Sectors', 33 *American Journal of International Law* 519–21 (1939).

Russia and Great Britain, 'Convention concernant les limites de leurs possessions respectives sur la côte Nord-Ouest de l'Amérique et la navigation de l'Océan Pacifique, conclue à Petersbourg 16/28 février 1825' (in French only), *Recueil De Martens*, N. S., Vol. II, at 427–8.

Russia and United States, 'Traité pour la cession de l'Amérique Russe aux Etats-Unis, signé à Washington le 30 mars 1857' (in French and in English), *Recueil De Martens*, N.R., 2e Série, Vol. I, at 39–40.

Smedal, G., *Acquisition of Sovereignty Over Polar Areas* 134 pages, at 54–76 (1931).

Smedal, G., *De l'acquisition de souveraineté sur les territoires polaires*, 208 pages, at 85–115 (1932).

Svarlien, O., 'The Legal Status of the Arctic', 52 *Proceedings of the American Society of International Law* 136–43 (1958).

Svarlien, O., 'The Sector Principle in Law and Practice', 10 *Polar Record* 248–63 (1960).

Triggs, G. D., *Australia's Sovereignty in Antarctica: the Validity of Australia's Claim at International Law*, Ph.D. dissertation at the University of Melbourne Law School, 673 pages, at 197–217 (1983).

United States (Department of State), 'U.S.–Russia Convention Line of 1867' *in International Boundary Study*, No. 14, at 1–3, with map showing boundary line (1 Oct. 1965).

Waldock, C. H. M., 'Disputed Sovereignty in the Falkland Islands', 25 *British Yearbook of International Law* 311–53 (1948).

Wall, E. H., 'The Polar Regions and International Law', 1 *International Law Quarterly* 54–8 (1947).

2 Historic waters

Affaire des Grisbadarna (1909), U.N. *Reports of International Arbitral Awards* Vol. XI, 147–66.

Baker Lake v. Minister of Indian Affairs and Northern Development (1980) 1 F.C. 518.

B.P. Exploration Co. (Libya) Ltd v. Hunt (1981) 1 W.W.R. 209.

Bernier, J. E., *Report of the Dominion Government Expedition to the Arctic Islands and the Hudson Strait on Board the D.G.S. 'Arctic', 1906–1907*, 127 pages (1909).

Bernier, J. E., *Report on the Dominion of Canada Government Expedition to the Arctic Islands and Hudson Strait on board the D.G.S. 'Arctic'* (1910).

Bernier, J. E., *Master Mariner and Arctic Explorer*, 409 pages (1939).

Blum, Y. Z., *Historical Titles in International Law*, 360 pages (1965).

Bourquin, M., Plaidoirie pour la Norvège, *Memorials, Anglo–Norwegian Fisheries Case*, Vol. IV, at 305–11 (1951).

Boyd, S. B., 'The Legal Status of Arctic Sea Ice', 22 *Canadian Yearbook of International Law* 98–152 (1984).

Calder et al. v. Attorney General of B.C. (1973) 4 W.W.R. 1.

Canada (Department of External Affairs), Historic Bays and Waters in 'Canadian Practice in International Law, 1973', 12 *Canadian Yearbook of International Law*, at 277–9 (1974).

Cooke, A. and Holland, C., 'Chronological List of Expeditions and Historical Events in Northern Canada 1821–45' 16 *Polar Record* 41–61 (1972).

Cooke, A., *Historical Evidence for Inuit Use of The Sea Ice*, paper delivered at Sikumiut Workshop, McGill University (15 April 1982).

Denhez, M., *Impact of Inuit Rights on Arctic Waters*, paper delivered at Sikumiut Workshop, McGill University (15 April 1982).

Dorion-Robitaille, Y., *Captain J. E. Bernier's Contribution to Canadian Sovereignty in the Arctic*, 110 pages (1978).

Dunbar, M. J., *The Biological Significance of Arctic Ice*, paper delivered at Sikumiut Workshop, McGill University (15 April 1982).

Freeman, M. (ed.), *Inuit Land Use and Occupancy Project*, 3 vols (1976).

Freeman, M., *Contemporary Inuit Exploitation of Sea Ice Environment*, paper delivered at Sikumiut Workshop, McGill University (15 April 1982).

Head, I. L., 'Canadian Claims to Territorial Sovereignty in the Arctic Regions', 9 *McGill Law Journal* 200–26 (1962–3).

Johnson, D. H. N., 'Acquisition Prescription in International Law', 27 *British Yearbook of International Law* 332–54 (1950).

Johnston, V. K., 'Canada's Title to the Arctic Islands', 14 *Canadian Historical Review* 24–41 (1933).

Konan, R. W., 'The *Manhattan*'s Arctic Conquest and Canada's Response in Legal Diplomacy', 3 *Cornell International Law Journal* 189–204 (1970).

Lester, G. S., *The Territorial Rights of the Inuit of the Canadian Northwest Territories: a Legal Argument*, J. D. dissertation at the Faculty of Law, York University, Toronto, 1530 pages (1981).

Loriot, F., *La théorie des eaux historiques et le régime juridique du Golfe Saint-Laurent en droit interne et international*, 705 typewritten pages (1972).

Low, A. P., *Report of the Dominion Government Expedition to the Hudson Bay and the*

Arctic Islands on Board the D.G.S. 'Neptune' 1903–4, 355 pages, particularly at 47–111 (1906).

MacGibbon, I. C., 'Customary International Law and Acquiescence', 33 *British Yearbook of International Law* 114–45 (1957).

Norway, 'Les titres historiques', Duplique de la Norvège, *Memorials, Anglo–Norwegian Fisheries Case,* Vol. III, at 438–43 (1951).

O'Connell, D. P., *The International Law of the Sea,* Vol. I, at 417–38 (1982).

Regina v. Tootalik E-321 (1969) 71 W.W.R. 435 and 74 W.W.R. 740.

Rivett-Carnac, C., 'The Establishment of the R.C.M.P. Presence in the Northwest Territories and the Arctic', 86 *Canadian Geographical Journal* 155–65 (1973).

Smith, G. W., *The Historical and Legal Background of Canada's Arctic Claims,* Ph.D. dissertation, Faculty of Political Science, Columbia University, 505 pages (1952).

Smith, G. W., 'The Transfer of Arctic Territories from Great Britain to Canada in 1880 . . .' 14 *Arctic* 54–73 (1961).

Smith, G. W., *Territorial Sovereignty in the Canadian North: A Historical Outline of the Problem,* NCRC-63-7, 13 pages (July 1963).

Smith, G. W., 'Sovereignty in the North: The Canadian Aspect of an International Problem', in R. St. J. Macdonald (ed.), *The Arctic Frontier,* at 194–255 (1966).

Smith, G. W., 'A Historical Summary of Maritime Exploration in the Canadian Arctic . . .' in *Proceedings of the International Commission of Maritime History,* 34 pages (Aug. 1970).

Smith, G. W., 'Canada's Arctic Archipelago', 27 *North/Nord,* No. 1, 10–15 and No. 2, 10–17 (1980).

Taylor, A., *Geographical Discovery and Exploration in the Queen Elizabeth Islands* 172 pages (1955).

United Nations (Secretariat), 'Juridical Regime of Historic Waters, including Historic Bays', A/CN.4/143 (9 March 1962), *Yearbook of the International Law Commission,* Vol. II, 1–26 (1962).

United Nations (Secretariat), *Historic Bays* A/CONF.13/1, (30 Sept. 1957).

Vallaux, G., 'Droits et prétentions politiques sur les régions polaires', *Affaires etrangeres* 14–33 (1932).

Waultrin, R., 'Le problème de la souveraineté des pôles', 15 *Revue génerale de Droit international public* 78–125, 185–209, 401–23, (1908).

Wenzel, G., *The Archaeological Evidence for Prehistoric and Inuit Use of the Sea Ice Environment,* paper delivered at Sikumiut Workshop, McGill University (15 April 1982).

3 Straight baselines

Alexander, L. M., 'Baseline Delimitations and Maritime Boundaries', 23 *Virginia Journal of International Law* 503–36 (1983).

Amerasinghe, C. F., 'The Problem of Archipelagos in the International Law of the Sea', 23 *International and Comparative Law Quarterly* 539–75 (1974).

Anand, R. P., 'Mid-Ocean Archipelagos in International Law: Theory and Practice', 19 *Indian Journal of International Law* 228–55 (1979).

Anglo–Norwegian Fisheries Case (1951) I.C.J. Rep. 4.

Beazley, B. P., 'Territorial Sea Baselines', 48 *International Hydrographic Review* 143–54 (1971).

Bishop, W. W., Jr., 'Fisheries Case', 46 *American Journal of International Law* 348–70 (1952).

Bourquin, M., 'La portée générale de l'Arrêt . . . dans l'Affaire anglo–norvégienne des pêcheries', 22 *Nordisk Tidsskrift for International Ret* 101–32 (1952).

Coquia, Jorge R., 'The Territorial Waters of Archipelagos', 1 *Philippine International Law Journal* 139–56 (1962).

Selected bibliography 273

Critchley, W. H., 'Polar Development of Soviet Submarines', 39 *International Journal* 828–65 (1984).
Dellapenna, J. W., 'Canadian Claims in Arctic Waters', 7 *Land and Water Law Review* 383–419 (1972).
Dubner, B. H., 'A Proposal for Accommodating the Interests of Archipelagic and Maritime States', 8 *New York University Journal of International Law* 39–61 (1975).
Dubner, B. H., *The Law of Territorial Waters of Mid-Ocean Archipelagos and Archipelagic States*, 115 pages (1976).
Evensen, J., 'The Anglo–Norwegian Fisheries Case and its Legal Consequences', 46 *American Journal of International Law* 609–30 (1952).
Evensen, J., *Certain legal aspects concerning the delimitation of the territorial waters of Archipelagos* Preparatory document No. 15, United Nations Doc. A/Conf.13/18 (29 Nov. 1957).
Feliciano, F. P., 'Comments on Territorial Waters of Archipelagos', 1 *Philippine International Law Journal* 157–77 (1962).
Goldie, L. F. E., 'The Archipelago Theory and the Third United Nations Conference on the Law of the Sea' *in Report of the Fifty-Eighth Conference, International Law Association*, 290–309 (1978).
Green, L. C., 'Canada and Arctic Sovereignty', 48 *Canadian Bar Review* 740–75 (1970).
Head, I. L., 'Canadian Claims to Territorial Sovereignty in the Arctic Regions', 9 *McGill Law Journal* 200–26 (1962–63).
Herman, L. L., 'Proof of Offshore Territorial Claims', 7 *Dalhousie Law Journal* 3–38 (1982).
Hodgson, R. D. and Alexander, L., *Towards an Objective Analysis of Special Circumstances, Occasional Paper*, No. 13, Law of the Sea Institute (April 1972).
Inch, D. R., 'An Examination of Canada's Claim to Sovereignty in the Arctic', 1 *Manitoba Law School Journal* 31–53 (1962).
Johnson, D. H. N., 'The Anglo–Norwegian Fisheries Case', 1 *International and Comparative Law Quarterly* 145–80 (1952).
Kobayashi, T., *The Anglo–Norwegian Fisheries Case of 1951 and the Changing Law of the Territorial Sea*, 87 pages (1965).
Konan, R. W., 'The *Manhattan*'s Arctic Conquest and Canada's Response in Legal Diplomacy', 3 *Cornell International Law Journal* 189–204 (1970).
Leversen, M. A., 'The Problems of Delimitation of Baselines for Outlying Archipelagos', 9 *San Diego Law Revue* 733–45 (1972).
Malek, C., 'La théorie dite des "baies historiques" ', 6 *Revue de droit international pour le Moyen-Orient* 100–73 (1957).
Marston, G., 'International Law and "Mid-Ocean" Archipelagos' 4 *Annales d'Etudes internationales* 171–90 (1973).
McConnell, W. H., 'The Legal Regime of Archipelagos', 35 *Saskatchewan Law Review* 121–45 (1970–1).
McRae, D. M., 'Arctic Waters and Canadian Sovereignty', 38 *International Journal* 476–92 (1983).
Morin, J. Y., 'La zone de pêche exclusive du Canada', 2 *Canadian Yearbook of International Law* 77–106 (1964).
O'Connell, D. P., 'Mid-Ocean Archipelagos in International Law', 45 *British Yearbook of International Law* 1–77 (1973).
O'Connell, D. P., *The International Law of the Sea*, Vol. I, at 171–235 (1982).
Parfond, P., 'L'Affaire anglo-norvégienne des pêcheries et le jugement de la Cour de Justice de La Haye du 18 décembre 1951', *Revue Maritime* 346–58 (1952).
Reid, R. S., 'The Canadian Claim to Sovereignty over the Waters of the Arctic', *Canadian Yearbook of International Law* III-136 (1974).

Reinhard, W. G., 'International Law: Implications of the Opening of the Northwest Passage', 74 *Dickinson Law Revue* 678–90 (1970).

Shinkaretskaya, G., *Sovereign States with Archipelagos* (in Russian); Moscow: Institute of State and Law, 102 pages (1977).

Smith, H. A., 'The Anglo–Norwegian Fisheries Case', *Year Book of World Affairs* 283–307 (1953).

Sorensen, M., 'The Territorial Sea of Archipelagos', 6 *Netherlands International Law Revue* 315–31 (1959).

United Nations, 'Report of the International Commission to the General Assembly', Doc. A/3159, *Yearbook of the International Law Commission*, Vol. II, at 267–8 (1956).

United Nations, Comments by Governments on Draft Articles by the International Law Commission, Doc. A/CONF.13/5 and Add. 1 to 4, *Official Records*, Vol. I, Preparatory Document No. 5, 75–113 (1958).

United Nations, *Convention on the Territorial Sea and the Contiguous Zone*, A/CONF.13/L.52 (1958).

United Nations, *Convention on the Law of the Sea Part II*, A/Conf.62/122 (7 Oct. 1982).

United States (Department of State), 'Sovereignty of the Sea', *Geographic Bulletin*, No. 3, Appendix A 'The Baseline', at 10–13 (April 1965).

Vanderzwaag, D. and Pharand, D., 'Inuit and the Ice: Implications for Canadian Arctic Waters', 21 *Canadian Yearbook of International Law* 53–84 (1983).

Waldock, C. H. M., 'The Anglo–Norwegian Fisheries Case', 28 *British Yearbook of International law* 114–71 (1951).

Wilberforce, R. O., 'Some Aspects of the Anglo–Norwegian Fisheries Case', *Transactions of the Grotius Society for 1952*, 151–68 (1953).

4 Northwest Passage

For a more complete bibliography, see this writer's book: *The Northwest Passage: Arctic Straits*, Martinus Nijhoff Publishers, 236 pages, at 186–92 (1984).

Anand, R. P., 'Freedom of Navigation through Territorial Waters and International Straits', 14 *Indian Journal of International Law* 169–89 (1974).

Anderson, W. R., 'The Arcticas and Sea Routes of the Future', *National Geographic Magazine* 21–4 (1959).

Arctic Institute of North America, *Arctic Marine Commerce Study* (1973): Vol. I, 305 pages; Vol II, 149 pages, prepared for Maritime Administration, U.S. Department of Commerce.

Baxter, R. R., *The Law of International Waterways*, 371 pages (1964).

Bohin, P., *Le passage des navires dans l'Océan Arctique*, mémoire de D.E.S.S., Faculté de droit, Brest, France, 231 pages (1984).

Bruël, E., *International Straits* (1945): Vol. I, The General Legal Position of International Straits, 278 pages; Vol. II, Straits Comprised by Positive Regulations, 426 pages.

Burke, W., 'Submerged Passage Through Straits: Interpretations of the Proposed Law of the Sea Treaty Text', 52 *Washington Law Review* 193–220 (1977).

Butler, Wm. E., *Northeast Arctic Passage*, 199 pages (1978).

Byrne, J., 'Canada and the Legal Status of Ocean Space in the Canadian Arctic Archipelago', 28 *University of Toronto Faculty of Law Revue* 1–16 (1970).

Canada, Department of Energy, Mines and Resources, *The National Energy Program*, 1980, 115 pages (1980).

Canada (Canadian Hydrographic Service), *Sailing Directions Arctic Canada*, 3rd ed., Vol. I (1982), Vol. II (1978), Vol. III (1981).

Canada (Dept. of Ind. Aff. & North. Dev.), *The Lancaster Region*, Draft Green Paper, 113 pages (1980).

Canadian Arctic Resources Committee, *Marine Transportation and High Arctic*

Development: Policy Framework and Priorities – Symposium Proceedings, 271 pages (1979).

Clingan, T. A., and Alexander, L. M., (eds.) *Hazards of Maritime Transit*, 140 pages (1973).

Corfu Channel Case Merits, (1949) I.C.J. Rep. 4.

Couper, A. (ed.), *The Times Atlas of the Oceans*, 272 pages (1983), at 62–3 (the Polar Oceans) and 148–9 (Seasonal Shipping Routes).

Critchley, H., 'Canadian Security Policy in the Arctic: the Context for the Future' in Canadian Arctic Resources Committee, *Marine Transportation and High Arctic Development: Policy Framework and Priorities*, at 181–209 (1979).

Critchley, H., 'Polar Deployment of Soviet Submarines', 39 *International Journal* 828–65 (1984).

Dugosevic, D., 'Le régime des détroits qui servent à la navigation internationale . . .' *Jugoslovenska Revija Za Medunarodno Pravo* 68–78 (1979).

Fillmore, S. and Sandilands, R. W., *The Chartmakers, the History of Nautical Surveying in Canada*, 225 pages (1983).

Frank, E. J., 'The Third United Nations Law of the Sea Convention and Straits Passage: the Maritime Powers Perspective on Transit Passage', 3 *Journal of International and Comparative Law* 243–70 (1982).

Gagne, M., 'The Legal Status of the Water Met by the "Manhattan" During her Voyage Through the Arctic', 11 *Les Cahiers de Droit* 66–73 (1970).

Griffiths, F., *A Northern Foreign Policy*, Canadian Institute of International Affairs, Toronto, 90 pages (1979).

Hill, C. E., 'Le régime international des détroits maritimes', 45 *Recueil des Cours*, Tome II, 475–556 (1934).

Howard, L. M. and Goodwin, C. R., *Beaufort Sea–Mackenzie Delta Environmental Impact Statement Bibliography*, 66 pages (1983).

Hume, H., *A Comparative Study of the Northern Sea Route and the Northwest Passage*, M.Phil. thesis, Scott Polar Research Institute, University of Cambridge, 107 pages (1984).

Hussain, I., 'The Law of the Sea Convention: the Right of Free Passage in Straits', 6 *Strategic Studies* 41–56 (1982).

Johnston, D. M. (ed.), *Arctic Ocean Issues in the 1980s*, 60 pages (1982).

Jones, J. M., 'The Corfu Channel Case: Merits', 26 *British Year Book of International Law* 447–53 (1949).

Khoshkish, A., *The Right of Innocent Passage*, 168 pages (1954).

Koh, K. L., *Straits of International Navigation*, 225 pages (1982).

Kuribayashi, T., 'The Basic Structure of the New Regime of Passage Through International Straits . . .', 21 *Japanese Annual of International Law* 29–47 (1977).

Labrousse, H., 'Les détroits après la Convention', in Société française pour le Droit international', *Perspectives du droit de la mer a l'issue de la 3e Conference des Nations Unies*, 29 pages (1983).

Lapidoth, R., *Les détroits en droit international*, 135 pages (1972).

Lapidoth, R., 'Freedom of Navigation: its legal History and its Normative Basis', 6 *Journal of Maritime Law and Commerce* 259–272 (1975).

Larsen, S., 'Passage Through Straits', 47 *Nordisk Tidsskrift for International Ret* 93–119 (1978).

Maduro, M. F., 'Passage through International Straits: The Prospects emerging from the Third United Nations Conference on the Law of the Sea', 12 *Journal of Maritime Law and Commerce* 65–95 (1980).

Marine Transportation Research Board (U.S.), *Maritime Services to Support Polar Resource Development*, Washington: National Academy Press, 78 pages (1981).

McConchie, R. D., and Reid, R. S., 'Canadian Foreign Policy and International Straits', in B. Johnson and M. W. Zacher (eds.), *Canadian Foreign Policy and the Law of the Sea*, 387 pages, at 158–201 (1977).

McLaren, A. S., 'Under the Ice in Submarines', 197 *U.S. Naval Proceedings* 105–9 (1981).
McLaren, A. S., 'The Development of Cargo Submarines for Polar Use', 21 *Polar Record* 369–81 (1983).
McNees, R. B., 'Freedom of Transit through International Straits', 6 *Journal of Maritime Law and Commerce* 175–211 (1975).
Momtaz, D., 'La Question des détroits à la Troisième Conférence des Nations-Unies sur le Droit de la mer', 20 *Annuaire Français de Droit international* 841–59 (1974).
Moore, John N., 'The Regime of Straits and the Third United Nations Conference on the Law of the Sea', 74 *American Journal of International Law* 77–121 (1980).
Nolta, F., 'Passage through International Straits: Free or Innocent? The Interests at Stake', 11 *San Diego Law Review* 815–33 (1974).
O'Connell, D. P., *The Influence of Law on Sea Power*, 204 pages (1975).
O'Connell, D. P., *The International Law of the Sea*, Vol I, at 260–337 (1982).
Pharand, D., 'Innocent Passage in the Arctic', 6 *Canadian Yearbook of International Law* 3–60 (1968).
Pharand, D., 'International Straits', 7 *Thesaurus Acroasium* 64–100 (1977).
Pharand, D., 'The Northwest Passage in International Law', 17 *Canadian Yearbook of International Law* 99–133 (1979).
Pharand, D., 'La contribution du Canada au développement du droit international pour la protection du milieu marin: le cas spécial de l'Arctique', 11 *Etudes Internationales* 441–66 (1980).
Pharand, D., *The Northwest Passage: Arctic Straits*, Martinus Nijhoff Publishers, 236 pages (1984).
Przetacznik, F., 'Freedom of Navigation Through Territorial Sea and International Straits', 55 *Revue de Droit international de Sciences diplomatiques et politiques* 222–36 and 299–319 (1977).
Pullen, T. C., 'A 3,000-mile Arctic Towing Odyssey', 101 *Canadian Geographic* 16–20 (Dec. 1981/Jan. 1982).
Pullen, T. C., 'Cruise Ship Completes Epic Voyage through Northwest Passage', 104 *Canadian Geographic* 20–29 (Dec. 1984/Jan. 1985).
Reddish, M. R., 'The Right of Passage by Warships Through International Straits', *Judge Advocate General Journal* 79–85 (1970).
Reinhard, W. G., 'International Law: Implications of the Opening of the Northwest Passage', 74 *Dickinson Law Review* 678–90 (1970).
Reisman, W. M., 'The Regime of Straits and National Security: An Appraisal of International Lawmaking', 74 *American Journal of International Law* 48–76 (1980).
Reisman, W. M., 'Inadequacies of the Straits' Passage Regime in the L.O.S. Draft', 5 *Marine Policy* 276–7 (1981).
Robertson, H. B., 'Passage Through International Straits: A Right Preserved in the Third United Nations Conference on the Law of the Sea', 20 *Virginia Journal of International Law* 801–57 (1980).
Roots, E. F., 'Environmental Aspects of Arctic Marine Transportation and Development' in Canadian Arctic Resources Committee, *Marine Transportation and High Arctic Development: Policy Framework and Priorities*, at 69–92 (1979).
Smith, B., 'Innocent Passage as a Rule of Decision: Navigation vs Environmental Protection' 21 *Columbia Journal of Transnational Law* 49–102 (1982).
Smith, R. W., 'Ocean Borne Shipment of Petroleum and the Impact of Straits on VLCC Transit', 1 *Maritime Studies and Management* 199–130 (1973).
Smith, R. W., 'Analysis of the Strategic Attributes of International Straits: A Geographical Perspective', 2 *Maritime Studies and Management* 88–101 (1974).
Steele, G. P., *Seadragon, Northwest Under the Ice*, 255 pages (1962).
Strong, J. T., 'The Opening of the Arctic Ocean', 87 *U.S. Naval Institute Proceedings* 58–65 (1961).

Tracey, N., 'Matching Canada's Navy to its Foreign Policy Requirements', 38 *International Journal* 459–75 (1983).

Treshnikov, A. F., *Mikhail Lomonosov and the Northern Sea Route*, 14 typewritten pages, paper presented at the centennial meeting of the Royal Society of Canada, in Ottawa, 3 June 1982.

Trudeau-Bedard, N., 'Souveraineté et Passage du Nord-Ouest', 5 *Revue Juridique Themis* 47–83 (1970).

Soldovnikoff, P., *La navigation maritime dans la doctrine et la pratique soviétiques*, 389 pages (1980).

United Nations, *Convention on the Law of the Sea*, Part III, A/Conf.62/122 (7 Oct. 1982).

Van der Mensbrugghe, 'Les canaux et détroits dans le droit de la mer actuel', in Rousseau C. and Weil, P. (eds), *Droit de la mer*, 256 pages, at 181–247 (1977).

Walker, P. B., 'What is Innocent Passage?' in Lillich, R. B. and Moore, J. N., *Readings in International Law* 365–87 (1980).

Westermeyer, W. E., and Shusterich, K. M., *United States Arctic Interests*, Springer-Verlag, 369 pages (1984).

Index

II Name index

III Geographical index

IV Ship index